PETROLEUM PRODUCTS

Applied Energy Technology Series

James G. Speight, Ph.D., *Editor*

Mushrush and Speight **Petroleum Products: Instability and Incompatibility**

IN PREPARATION

Khan **Conversion and Utilization of Waste Materials**

PETROLEUM PRODUCTS:
INSTABILITY AND INCOMPATIBILITY

George W. Mushrush, Ph.D.
Chemistry Department
George Mason University
4400 University Drive
Fairfax, Virginia 22030-4444
USA

and

James G. Speight, Ph.D.
Western Research Institute
365 North 9th Street
Laramie, Wyoming 82070-3380
USA

CRC Press
Taylor & Francis Group
Boca Raton London New York

CRC Press is an imprint of the
Taylor & Francis Group, an **informa** business
A TAYLOR & FRANCIS BOOK

First published 1995 by Taylor & Francis

Published 2018 by CRC Press
Taylor & Francis Group
6000 Broken Sound Parkway NW, Suite 300
Boca Raton, FL 33487-2742

PETROLEUM PRODUCTS: Instability and Incompatibility

© 1995 by Taylor & Francis Group, LLC
CRC Press is an imprint of Taylor & Francis Group, an Informa business

First issued in paperback 2019

No claim to original U.S. Government works

ISBN 13: 978-0-367-44899-8 (pbk)
ISBN 13: 978-1-56032-297-9 (hbk)

Visit the Taylor & Francis Web site at
http://www.taylorandfrancis.com

and the CRC Press Web site at
http://www.crcpress.com

This book was set in Courier by G. W. Mushrush. The editor was Chris Williams. Cover design by Michelle Fleitz.

A CIP catalog record for this book is available from the British Library.

Library of Congress Cataloging-in-Publication Data
Mushrush, George, W.
  Petroleum products: instability and incompatibility/ George W. Mushrush and James G. Speight.
    p. cm.
  Includes biographical references.

  1. Petroleum products. I. Speight, J. G. II. Title.
TP690.M833 1995
665.5'3—dc20
                                                          95-17340
                                                             CIP

CONTENTS

The emergence of petroleum as a plentiful and cheap source of energy has led to the evolution of a variety of liquid fuels and other products. Recent trends have shown that the "quality" of the petroleum is decreasing and refineries have to accommodate this change. One area where such an effect is particularly noticeable is in the character of the products.

Differences in crude oil composition and in the nature of the products from refinery operations have all served to emphasize the issue of fuel incompatibility. In fact, the incompatibility of petroleum and liquid fuels has become a major issue over the past two decades.

There have also been serious attempts to produce liquid fuels from the so-called "unconventional" sources such as coal, oil sands (often referred to as tar sands or bituminous sands), and oil shale. Indeed, this reemphasis of the value of these unconventional liquid fuel sources has also caused the issue of incompatibility to emerge.

Although the production of various products from sources other than petroleum is still distant, instability and incompatibility are real.

This book is the outcome of studies by the authors into the various aspects of incompatibility exhibited, under certain circumstances, by liquid fuels and other products.

In more general terms, the book is written as a teaching text from which the reader can gain a broad overview (with some degree of detail), of the chemical and physical concepts of instability and incompatibility.

The nature of the subject virtually dictates that the text include some chemistry (and the present text is not delinquent in this respect), but attempts have been made for the benefit of those readers without any formal post-high school training in chemistry (and who may, therefore, find chemistry lacking in any form of inspiration) to maintain the chemical sections in the simplest possible form. Wherever possible, simple chemical formulas have been employed to illustrate the text.

The text will, therefore, satisfy those who are just entering into this fascinating aspect of science and engineering as well as those (scientists and engineers) who are already working with liquid fuels but whose work is so specific that they also require a general overview. It will also be of assistance to petroleum refinery personnel, who may one day be called upon to handle liquid fuels from other sources as feedstocks for the refinery system.

References are cited throughout the text and have been

selected so that the reader might use the citations for more detail. It would have been impossible to include all of the relevant references for a subject of such diverse character. Thus, where possible, references such as review articles, other books, and those technical articles with substantial introductory material have been used in order to pass on the most information to the reader. The reader might also be surprised at the number of older references that are included. The purpose of this is to remind the reader that there is much valuable work cited in the older literature. This is particularly true of some of the older chemical concepts, where many of the ideas are still pertinent, and we should not forget these valuable contributions to fuel science and technology.

For the benefit of those readers who have had formal training in one (or more) of the engineering disciplines, the text contains both the metric and nonmetric measures of temperature (Celsius and Fahrenheit). However, it should be noted that exact conversion of the two scales is not often possible and, accordingly, the two temperature scales are converted to the nearest 5°. At the high temperatures often quoted in the process sections, serious error will not arise from such a conversion.

With regard to the remaining metric/nonmetric scales of measurement, there are also attempts to indicate the alternate scales.

For the sake of simplicity and clarity, simple illustrations (often line drawings) are employed for the various process options, remembering, of course, that a line between two reactors may be not only a transfer pipe but also a myriad of valves and control equipment.

George Mushrush
James G. Speight

CHAPTER 1: INSTABILITY AND INCOMPATIBILITY

1.0 Introduction

The incompatibility of petroleum and liquid fuels has become a major issue over the past two decades. A refinery produces a wide variety of products (Figure 1.1) and is dependent upon the character of the crude oil (Speight, 1991; McKetta, 1992; Gray, 1994). Differences in crude oil composition and in the nature of the products from refinery operations have all served to emphasize the issue of fuel incompatibility.

However, before proceeding, there are several terms that need defining in order to understand the language of the chemist/engineer/chemical engineer, who study issues such as instability in petroleum and petroleum products.

The best means of approaching the issue of incompatibility is from an understanding of the nature of petroleum, the predominant source of liquid fuels, and the other sources of liquid fuels. In fact, it is essential to understand not only the nature of the liquid fuels themselves but also the chemical and physical nature of the liquid fuel sources.

This is not an attempt to justify assigning a major segment of this text to characterization studies of petroleum, coal, and oil shale. It is, more appropriately, a means of understanding the chemistry, physics, and behavior of liquid fuel precursors and, hence, the behavior of liquid fuels themselves.

By way of introduction, petroleum is a hydrocarbon fuel that contains additional elements such as nitrogen, oxygen, sulfur, and some metals. The atomic hydrogen/carbon ratio of petroleum varies but is generally of the order of 1.5-1.6 (Figure 1.2). Other sources of liquid fuels such coal and the bituminous material isolated from tar sand formation are somewhat lower. Thus, in order to produce liquid fuels from such materials, the atomic hydrogen/carbon ratio must be increased to ca. 1.8-2.0.

The classification of these resources is difficult and can only be shown in general terms (Figure 1.3). However, the definition of the fossil fuels as hydrocarbon resources is used liberally insofar as each resource contains significant amounts of nitrogen, oxygen, and sulfur.

Understanding this phenomenon is crucial to understanding the means by which instability and incompatibility can arise.

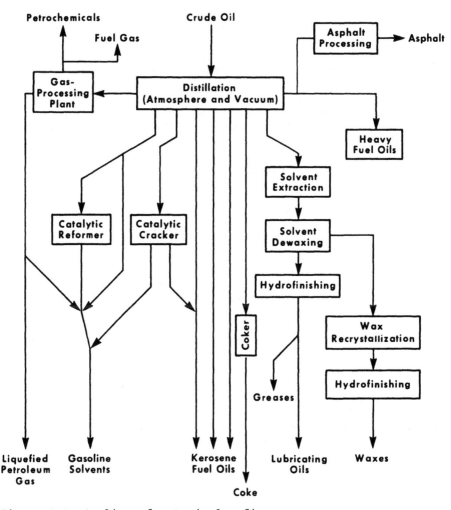

Figure 1.1: Outline of a typical refinery

2.0 Definitions

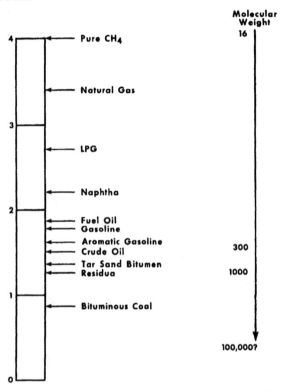

Figure 1.2: Representation of hydrogen content and molecular weight of fossil fuels.

Although the terminology applied to liquid fuels and liquid fuel sources has evolved over the decades since the first petroleum refinery in 1859, confusion still exists. In addition, there is always the possibility that other liquid fuel sources such as coal and oil shale (National Research Council (NRC), 1990) may emerge from the layers of darkness that have enshrouded them.

Therefore, it is appropriate to define here some of the terms that are used in the liquid fuels field so that their use later in

3

the text will be more apparent and will also alleviate some potential for misunderstanding.

2.1 General Terminology

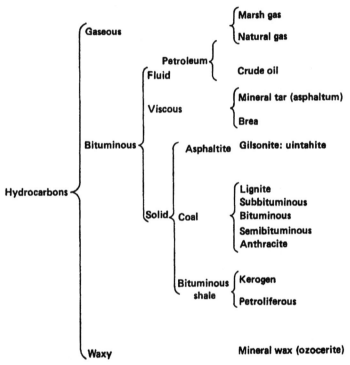

Figure 1.3: Simple classification scheme for hydrocarbon resources.

The general scientific areas of **fuel instability** and **fuel incompatibility** are complex and have been considered to be nothing better than a black art because not all of the reactions that contribute to instability and incompatibility have been defined (Wallace, 1964). Nevertheless, studies over the past three decades

have made valuable contributions to our understanding of instability and incompatibility in fuels. For the most part, however, gaps remain in our knowledge of the chemistry and physics of instability and incompatibility.

Briefly, **incompatibility** refers to the formation of a precipitate (or sediment) or phase separation when two liquids are mixed. The term **instability** is often used in reference to the formation of color, sediment, or gum in the liquid over a period of time. This term may be used in contrast to the formation of a precipitate in the near, almost immediate term. However, the terms are often used interchangeably.

Gum formation alludes to the formation of soluble organic (usually considered to be polymeric) material, whereas sediment is considered to be insoluble organic (usually polymeric) material. **Storage stability** (or **storage instability**) is a term used to describe the ability of the liquid to remain in storage over extended periods of time without appreciable deterioration as measured by gum formation and the depositions of insoluble material. **Thermal stability** is also defined as the ability of the liquid to withstand relatively high temperatures for short periods of time without the formation of carbonaceous deposits (coke) (Brinkman and White, 1981). There is also the tendency to define stability as the ability of the liquid to withstand long periods at temperatures up to 100°C (212°F) without degradation.

In general, fuel instability and incompatibility can be related to the existence of heteroatoms(i.e. containing nitrogen, oxygen, and sulfur). The degree of unsaturation of the fuel (i.e. the level of olefinic species) also plays a role in determining instability/incompatibility. Also, recent investigations have implicated catalytic levels of various oxidized intermediates and acids as especially deleterious for middle distillate fuels.

The phenomenon of **instability** (often variously referred to as **incompatibility** and more commonly known as **sludge formation** and **sediment formation** or **deposit formation**) in petroleum, liquid fuels, and other products often manifests itself in various ways (Stavinoha and Henry, 1981; Hardy and Wechter, 1990; Ruzicka and Nordenson, 1990; Power and Mathys, 1992). Hence, there are different ways of defining each of these terms.

For example, **instability** may be observed during crude oil storage, refining, product storage, and product utilization. The

5

composition of the sludge/sediment often provides some insights into the mode of formation.

Instability of petroleum, liquid fuels, and other products often manifests itself through the formation of an insoluble material, often referred to as **sludge**. The sludge may be solid, semisolid, or gellike. In addition, the sludge is often referred to as being **incompatible** with the surrounding liquid medium.

Instability reactions are usually defined in terms of the formation of filterable and nonfilterable sludge (sediments, deposits, and gums), an increased peroxide level, and the formation of color bodies. Color bodies in and of themselves do not predict instability; however, the reactions that initiate color body formation can be closely linked to heteroatom-containing (i.e., nitrogen-, oxygen-, and sulfur-containing) functional group chemistry.

Fuel **incompatibility** can have many meanings. The most obvious example of incompatibility (immiscibility) is the inability of a hydrocarbon fuel and water to mix. In the present context, incompatibility usually refers to the presence of various polar functionalities (i.e., heteroatom function groups containing nitrogen, oxygen, or sulfur, and even various combinations of the heteroatoms) in the crude oil.

As is often the case, some of the polar functional groups can survive the refining operations (perhaps with some change) and end up in the finished fuel products. Thereafter, the groups can participate in a variety of chemical reactions, leading to fuel deterioration and/or instability and incompatibility.

Perhaps the true meaning of the term **incompatibility** is found when it is applied to the refining industry. The application occurs when a product that is incompatible with (immiscible with or insoluble in) other products is formed in the same reactor. Such an example is the formation of coke during many of the thermal and catalytic operations. Coke formation is considered to be an initial **phase separation** of an insoluble, solid, coke precursor prior to coke formation proper.

In the case of crude oils, **sediments** and **deposits** are closely related to **sludge**, at least as far as compositions are concerned. The major difference appears to be in the character of the material.

By way of often archaic terminology, sludge that settles as a bottom layer in storage facilities is often semisolid, whereas a **sediment** and a **deposit** are usually solid materials. These are mere semantics, but are often very real differences to the operators of storage facilities.

There is also the suggestion (often, but not always, real) that the sediments and deposits originate from the inorganic constituents of petroleum. They may be formed from the inherent components of the crude oil (i.e., the metalloporphyrin constituents) or from the ingestion of contaminants by the crude oil during the initial processing operations. For example, crude oil is known to "pick up" iron and other metal contaminants from contact with pipelines and pumps.

Sediments can also be formed from organic materials, but the fact is that when reference is made to sediments and deposits, the inference is that these materials are usually formed from inorganic materials such as salts, sand, rust, and other contaminants that are insoluble in the crude oil and that settle to the bottom of the storage vessel.

The **sediment** and other degradation products can have a marked influence on the properties of finished fuels. Degradation products can affect color, volatility, flow characteristics of middle distillates, freezing point, combustion properties, and rheological properties of asphalt (Traxler, 1961; Barth, 1962; Hoiberg, 1964; Broome and Wadelinn, 1973).

Gum typically forms by way of a hydroperoxide intermediate that induces polymerization of olefins. The intermediates are usually soluble in the liquid medium. However, gums that have undergone extensive oxidation reactions tend to be higher in molecular weight and much less soluble.

Thus, the sediments of high molecular weight sediments that form in fuels are usually the direct result of autoxidation reactions. Active oxygen species involved include both molecular oxygen and hydroperoxides. These reactions proceed by a free radical mechanism, and the solids produced tend to have increased incorporations of heteroatoms and are, thus, also more polar and increasingly less fuel soluble.

Fuel instability problems can be of two related types: high-temperature short-term reactions (oxidative instability) and ambient temperature long-term reactions (storage instability). Oxygen, nitrogen, and sulfur species can be implicated in both types of instability. The synergism between organic sulfur and nitrogen compounds with active oxygen species on the stability of middle distillate fuels is not well understood. Differences in fuel composition, reaction surface, quantity of dissolved oxygen, hydroperoxide concentration, and reaction temperature all contribute to the reported variations in fuel deterioration.

The most significant and undesirable instability change in fuel liquids is the formation of solids, termed **filterable sediment**. Filterable sediments can plug nozzles and filters, coat heat exchanger surfaces, and otherwise degrade engine performance. These solids are the result of free radical autoxidation reactions. Although slight thermal degradation occurs in nonoxidizing atmospheres, the presence of oxygen or active oxygen species, i.e., hydroperoxides, will greatly accelerate oxidative degradation as well as significantly lower the temperature at which undesirable products are formed. Solid deposits that form as the result of short-term high-temperature reactions share many similar chemical characteristics with filterable sediments that form in storage. Fuel instability is consequently dependent on the nature of potential autoxidation pathways.

The soluble sludge/sediment precursors that form during processing or use may have a molecular mass in the several hundreds range. For this soluble precursor to reach a molecular weight sufficient to precipitate (or to phase-separate), one of two additional reactions must occur. Either the molecular weight must increase drastically as a result of condensation reactions, leading to the higher molecular weight species or the polarity of the precursor must increase (without necessarily increasing the molecular weight) by incorporation of additional oxygen, sulfur, or nitrogen functional groups. In both cases, insoluble material will form and separate from the liquid medium.

Both stability and incompatibility result in the formation and separation of degradation products or in other undesirable changes in the original properties of petroleum products. In the case of instability, there is a low resistance of the product to environmental influences during storage or to its susceptibility to oxidative or other degradation processes. In the case of incompatibility, degradation products form or changes occur due to an interaction of some chemical groups present in the components of the final blend.

Petroleum product components can be defined as being incompatible when sludge, semisolid, or solid particles are formed when such fuels are blended. The products so formed can be either soluble or insoluble in the final blend. In the former case, they usually increase the viscosity of the product; in the latter case, they settle out as semisolid or solid matter floating in the fuel or being deposited on the walls and floors of containers.

Incompatibility phenomena, although not always recognized as such, can be associated with most of the petroleum products. The phenomenon of incompatibility of petroleum, liquid fuels, and other products is invariably associated with the chemical composition of the components. In most cases a certain component in one of the fuels reacts with another component in the fuel with which it is blended. This chemical reaction results in the formation of a new product that, when soluble, affects the properties of the blend and, when insoluble, settles out in the form of semisolid or solid matter.

A special case regarding incompatibility of petroleum products is represented by **additives** incorporated into the various products. Additives are chemical compounds intended to improve some specific properties of fuels or other petroleum products. Different additives, even when added for identical purposes, may be incompatible with each other, for example they may react and form new compounds. Consequently, a blend of two or more fuels containing different additives may form a system in which the additives react with each other and so deprive the blend of any beneficial effects.

In general terms, studies of the composition of the incompatible materials often involve determination of the distribution of the organic components by **selective fractionation,** analogous to the deasphalting procedure (Speight, 1991):

(1) heptane soluble material, often called maltenes or petrolenes in petroleum work (Speight, 1991);

(2) heptane insoluble material and benzene (or toluene) soluble material, often referred to as asphaltenes;

(3) benzene (or toluene) insoluble material, referred to as carbenes and carboids when the fraction is a thermal product;

(4) pyridine soluble material (carbenes) and pyridine insoluble material (carboids).

Carbon disulfide and tetrahydrofuran have been used in place of pyridine. Carbon disulfide, although having an obnoxious odor and therefore not much different from pyridine, is easier to remove because of the higher volatility. Tetrahydrofuran is not as well established in petroleum research as it is in coal-liquid research. Thus, it is more than likely that the petroleum researcher will use carbon disulfide, pyridine, or some suitable alternate solvent.

It may also be necessary to substitute cyclohexane as an additional step for treatment of the heptane insoluble materials prior to treatment with benzene (or toluene). The use of quinoline has been suggested in place of pyridine, but this solvent presents issues associated with the high boiling point of the solvent.

Whichever solvent separation scheme is employed, there should be ample description of the procedure that the work can be repeated not only in the same laboratory by a different researcher but also in different laboratories by different researchers. Thus, for any particular feedstock and solvent separation scheme, the work should repeatable and reproducible within the limits of experimental error.

The sludge formed by the instability of crude oil (or by the incompatibility of the crude oil constituents) often contains inorganic as well as organic components, and sometimes both intermingled in a heterogeneous matrix. The inorganic components are represented often by aqueous solutions of salts, sand, corrosion products, bacterial slime, and other contaminants. The organic components in crude oil sludge are usually waxes, waxlike substances, polar constituents, or condensed high molecular weight

Table 1.1: Ranges of fractions usually found in sludge/sediment.

Fraction	Range (wt%)
Heptane solubles	0-10
Heptane insolubles	10-60
Toluene insolubles	10-25
Wax	0-10
Water	0-50

constituents (such as asphaltene material), and a variety of other
organic (degradation) products (Table 1.1) (Por, 1992). In
addition, there has been some success in determining the type of
functional groups by derivatization (Power and Mathys, 1992).
  The degradation products formed during the long term storage
of crude oils have been identified (Mochida et al., 1986) and it
has been postulated that the instability/incompatibility is a
result of air oxidation, photooxidation, and condensation. Any
high molecular weight polar substances that stay solid at ambient
temperatures are so formed and, consequently, form sludge.
  To summarize, part of the problem of incompatibility arises
from the manner in which the fuels (petroleum and its product
fractions, as well as other related liquids) react with oxygen
during storage. The remainder of the problem may be due to the
character of the individual constituents. Many of the constituents
may be incompatible/immiscible with each other, thereby causing
phase separation and hence, incompatibility.
  Another part of the problem (not to be underestimated) arises
because of variations in terminology. The nonspecific application
of various names is confusing and needs to be resolved. Part of
the goal of this chapter is to attempt to resolve the issue of
nomenclature. The terminology may seem confusing unless some

Table 1.2: Nomenclature and boiling ranges of the various
distillation fractions of petroleum

Fraction	Boiling Range	
	°C	°F
Light naphtha	-1 - 150	30 - 300
Heavy naphtha	150 - 205	300 - 400
Gasoline	-1 - 180	30 - 355
Kerosene	205 - 260	400 - 500
Stove oil	205 - 290	400 - 550
Light gas oil	260 - 315	400 - 600
Heavy gas oil	315 - 425	600 - 800
Lubricating oil	>400	>750
Vacuum gas oil	425 - 600	800 -1050
Residuum	>600	>1050

general rules are accepted. To alleviate some of the confusion, the terms **instability** and **incompatibility** will be used with caution throughout the text, and the meanings will be made clear at the time of use.

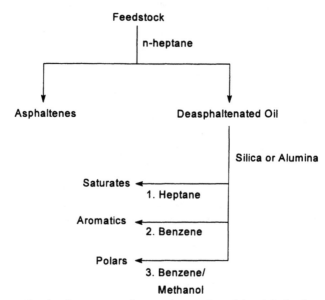

Figure 1.4: Typical separation scheme for liquid fuels

2.2 Petroleum Related Terms

Crude petroleum is a mixture of compounds boiling at different temperatures that can be separated into a variety of different generic fractions (Table 1.2). However, the general premise is that petroleum is a continuum of chemical types formed by maturation of various natural product precursors. Furthermore, regional variations in the amounts of the source materials included in the prepetroleum mixture as well as regional variations in the maturation conditions are the predominant causes for variations in petroleum (Speight, 1991).

In addition to separation by distillation, petroleum can also be separated into fractions based on solubility and on adsorption properties (Figure 1.4). As a result, it is possible to classify petroleum as paraffinic or naphthenic. Aromatic and asphaltic classifications also exist, but more as subcategories of the naphthenic classification.

Thus, the nonasphaltic (volatile) portion of petroleum has been "mapped" to show the distribution of the various compound classes (Figure 1.5). However, it must be recognized that each petroleum portion would show a different distribution and therefore require different treatment to produce a given slate of products (Speight, 1991).

Since there is a wide variation in the properties of crude petroleum (Speight, 1991), the proportions in which the different fractions occur vary with origin. Thus, some crude oils have higher proportions of the lower-boiling components, while others have higher proportions of residuum (asphaltic components).

Petroleum, and the equivalent term **crude oil**, covers a vast assortment of materials consisting of gaseous, liquid, and solid hydrocarbon-type chemical compounds that occur in sedimentary deposits throughout the world (Speight, 1991). Indeed, there are many terms that are related (Table 1.3) and need to be understood to alleviate the possibility of nomenclatorial confusion!

The constituents of petroleum vary widely in specific gravity and viscosity. Metal-containing constituents, notably those compounds that contain vanadium and nickel, usually occur in the more viscous crude oils in amounts up to several thousand parts per million and can affect the processing of these feedstocks (Speight, 1990).

When petroleum occurs in a reservoir that allows the crude material to be recovered by pumping operations as a free-flowing, low-density (specific gravity), dark- to light-colored liquid, it is often referred to as **conventional petroleum** (Table 1.4).

It is worthy of note here that there is the potential for incompatibility to occur in the reservoir. The increase in reservoir pressure due to the generation of gases during maturation will cause some of these gases to dissolve in the fluid. This causes changes in the composition of the reservoir fluid (Evans et al., 1971), and the consequences are analogous to the deasphalting process (Figure 1.8), in which the asphaltic constituents (mostly the polar resins and the asphaltenes) are rendered incompatible (insoluble) (Figure 1.6) and are deposited on the reservoir rock.

13

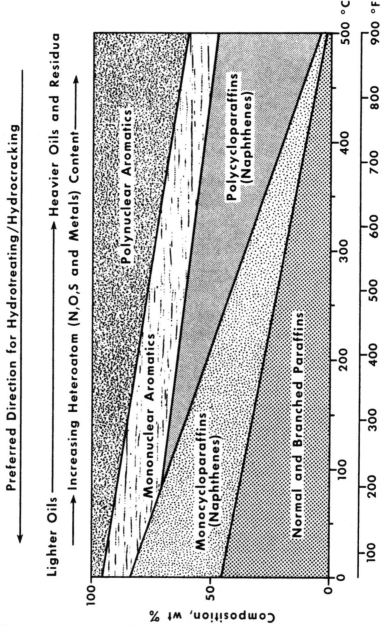

Figure 1.5: Crude oil composition

Table 1.3: Petroleum and related materials can be divided into various subgroups.

Natural Materials	Manufactured Materials	Derived Materials
Petroleum	Wax	Oils
Heavy Oil	Residuum	Resins
Mineral Wax	Asphalt	Asphaltenes
Bitumen	Tar	Carbenes
Asphaltite	Pitch	Carboids
Asphaltoid	Coke	
Migrabitumen	Synthetic Crude Oil	
Bituminous Rock		
Bitumenous Sand		
Kerogen		
Natural Gas		

Footnotes:
1. Tar sand: a misnomer, tar is a product of coal processing.
2. Oil Sand: also a misnomer but equivalent to usage of "oil shale."
3. Bituminous sand: more correct.
4. Asphalt: a product of a refinery operation; usually made from a residuum.
5. Bitumen: often referred to as "native asphalt."
6. Residuum: the nonvolatile portion of petroleum and often further defined as "atmospheric" (b.p. >350°C; >660°F) or vacuum (b.p. >565°C; >1050°F).

The converse can also be true when the reservoir is in production and the gases leave the solution. There is the potential for the fluid to redissolve the asphaltic constituents. **Heavy oil** is another type of petroleum that is different from conventional petroleum in that the high viscosity (Table 1.4; Figure 1.7) translates to reduced flow properties. A heavy oil is much more difficult to recover from the subsurface reservoir. Recovery of this type of petroleum usually requires thermal stimulation of the reservoir.

On this point, it is worthy of note that a viscosity increase is usually accompanied by increases in the amount of nitrogen, sulfur, and carbon residue (Figure 1.8). The indications are that heavy oils are more difficult to refine than conventional petroleum and can afford a different slate of products. Such differences can cause instability in the products as well as incompatibility during refining operations.

Thus, the generic term **heavy oil** is often applied to a petroleum that has an American Petroleum Institute (API) gravity of less than 20° and usually, but not always, a sulfur content higher than 2% by weight. Furthermore, in contrast to conventional crude oils, heavy oils are darker in color and may even be black.

The term **heavy oil** has also been arbitrarily used to describe both the heavy oils that require thermal stimulation of recovery from the reservoir and the bitumen in bituminous sand (tar sand, oil sand) formations from which the heavy bituminous material is recovered by a mining operation.

Extra heavy oil is that petroleum related material that occurs in the near-solid state and is virtually incapable of free flow under ambient conditions.

Bitumen (often referred to as **native asphalt**) (Table 1.3) is

Table 1.4   Classification of petroleum by density-gravity.

Type of crude	Characteristics
1. Conventional/"light" crude oil	Density <934 kg/gm^3 (>200°API)
2. "Heavy" crude oil	Density:1000 kg/m^3 >934 kg/m^3 100° API to <20° API) Maximum viscosity: 10,000 mPa (cp)
3. "Extra-heavy" crude oil; may also include atmospheric residua	Density >1000 kg/m^3 (<10° API) Maximum viscosity: 10,000 mPa (cp) (b.p.> 340°C; > 650°F)
4. Bitumen may also include vacuum residua	Viscosity >10,000 mPa (cp) Density >1000 kg/m^3 <10°API (b.p. > 510°C;>950°F)

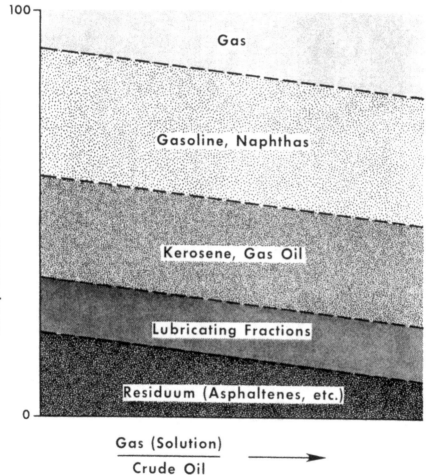

Figure 1.6: Effect of gas/oil ratio on composition.

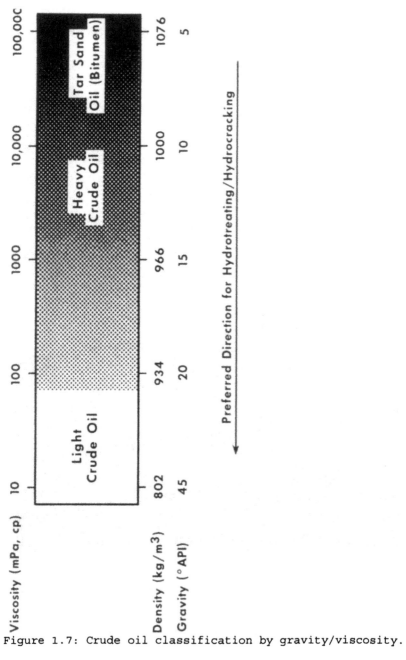

Figure 1.7: Crude oil classification by gravity/viscosity.

Table 1.5: Typical properties of bitumen and crude oil.

Property	Bitumen	Crude oil
Elemental analysis, wt%		
carbon	83	86
hydrogen	11	13
nitrogen	0.5	0.2
oxygen	1	0
sulfur	5	<2
Gravity, °API	<10	>25
Viscosity		
sus, 100°F (38°C)	35,000	<30
Pour point (°F)	50	< 0
Asphaltenes	15	< 5
Metals, ppm		
vanadium	250	<50
nickel	100	<25
Carbon residue		
Conradson	15	< 2

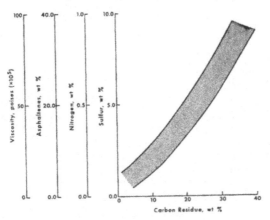

Figure 1.8: Variation of carbon residue with other properties.

a subclass of extra heavy oil and includes a wide variety of reddish brown to black materials of semisolid, viscous to brittle character that can exist in nature with no mineral impurity or with mineral matter contents that exceed 50% by weight.

Bitumen is more viscous than conventional crude oil (Table 1.5), is much less volatile (Table 1.6), and has a higher proportion of asphaltenes and resins (Table 1.7). Bitumen is frequently found as the organic filling in pores and crevices of sandstones, limestones, or argillaceous sediments, in which case the organic and associated mineral matrix is known as rock asphalt.

The expression **tar sand** is commonly used in the petroleum industry to describe a sandstone reservoir that is impregnated with a viscous black, extra heavy crude oil. This type of petroleum cannot be retrieved through a well by conventional production techniques. However, the term **tar sand** is actually a misnomer. It is incorrect to refer to native bituminous materials as **tar** or **pitch**. Although the word "tar" is descriptive of the black, heavy bituminous material, it is best to avoid its use with respect to natural materials. Alternative names, such as **bituminous sand** or

Table 1.6: Distillation profiles for bitumen and crude oil.

Cut point		Distillate	
°C	°F	wt%	
		Bitumen	Crude oil
200	390	0 - 3	35
225	435	1 - 5	40
250	480	2 - 7	45
275	525	4 - 9	51
300	570	8 - 14	
325	615	9 - 26	
350	660	10 - 28	
400	750	15 - 30	
450	840	20 - 33	
500	930	25 - 40	
540	1000	30 - 45	
540+		55 - 70	

Table 1.7: Fractional composition of bitumen.

Fraction	Range wt%
Asphaltenes	6-40
Resins	20-40
Aromatics	20-35
Saturates	20-35

oil sand, are gradually finding usage, the former name (bituminous sands) being more technically correct. The term "oil sand" is used in the same way as the term "tar sand," and these terms are often used interchangeably (Speight, 1990). More correctly, the name tar is usually applied to the high-boiling product remaining after the destructive distillation of coal or other organic matter. The meaning should be restricted to the volatile or near-volatile products produced in the destructive distillation of such organic substances as coal. In the simplest sense, pitch is the distillation residue of the various types of tars.

A residuum (also shortened to "resid") is the residue obtained

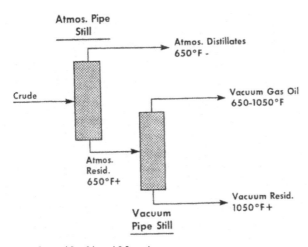

Figure 1.9: Crude oil distillation.

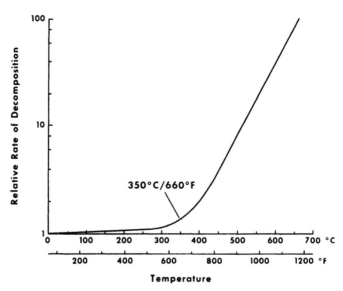

Figure 1.10: Distillation profile showing decomposition temperature.

Table 1.8: Properties of residua (Tia Juana light crude oil.

Property			Cut Point		
		°F	650	950	1050
		°C	345	510	565
Yield, wt%	100		49	24	18
Gravity, °API	32		17	10	7
Sulfur, wt%	1		1.8	2.4	2.6
Carbon residue,					
Conredson, wt%			9	17	22
Pour point, °F	-5		45	95	120
Metals, ppm					
Vanadium	<50		185		450
Nickel	<10		25		64

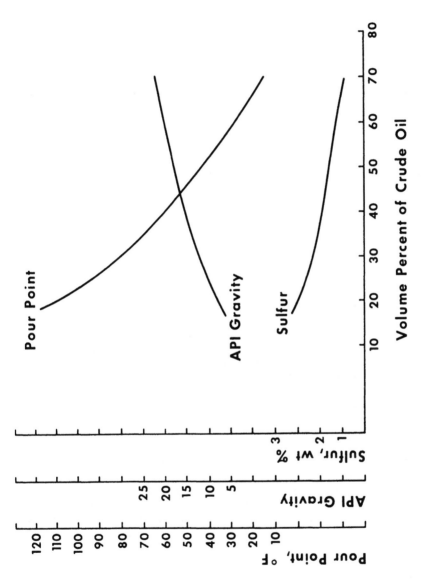

Figure 1.11: Variation of residua properties with cut point.

Table 1.9:   Petroleum classification according to chemical
              composition.

Class of crude	Composition of 250-300°C fraction, %				
	Paraffins	Naphthenes	Aromatics	Wax	Asphalt
Paraffinic	46-61	22-32	12-25	1-10	0- 6
Paraffinic- naphthenic	42-45	38-39	16-20	1- 6	0- 6
Naphthenic	15-26	61-76	8-13	trace	0- 6
Paraffinic- naphthenic- aromatic	27-35	36-47	26-33	0- 1	0-10
Aromatic	0- 8	57-78	20-25	0- 1	0-20

from petroleum after nondestructive distillation has removed all
the volatile materials (Figure 1.9). The temperature of the
distillation is usually maintained below 350°C (660°F), since the
rate of thermal decomposition of petroleum constituents is
substantial above 350°C (660°F) (Figure 1.10).

Residua (plural of **residuum**) are black, viscous materials and
are obtained by distillation of a crude oil under atmospheric
pressure (atmospheric residuum) or under reduced pressure (vacuum
residuum). They may be liquid at room temperature (generally
atmospheric residua) or almost solid (generally vacuum residua),
depending upon the nature of the crude oil and the temperature of
the distillation (Table 1.8; Figures 1.11 and 1.12). In addition,
different crude oils produce different amounts of residua (Figure
1.13), and the properties are not guaranteed to be the same. The
differences between a parent petroleum and the residua are due to
the relative amounts of various constituents present, which are
removed or remain by virtue of their relative volatility.

When a residuum is obtained from a crude oil and thermal
decomposition has occurred (such as in various thermal operations,
of which visbreaking might be an example), it is usual to refer to
this product as **pitch**. By inference, the name is used in the same

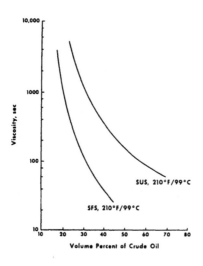

Figure 1.12: Variation of crude oil properties with cut point.

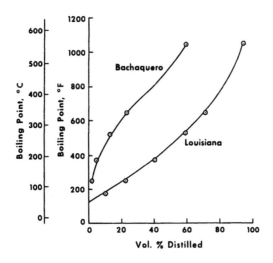

Figure 1.13: Variation of crude oil properties with cut point.

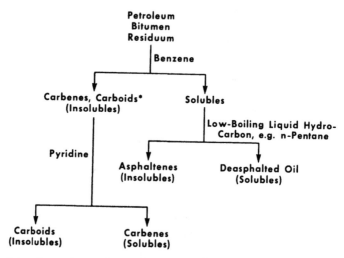

Figure 1.14: Fractionation of crude oil.

manner as when it is used to refer to the nonvolatile residue from the thermal decomposition of coal tar. However, the chemical composition of **petroleum pitch** (**vacuum tower pitch**, etc.) is generally quite different from **coal tar pitch**.

The chemical composition of crude oil varies, depending upon the nature of the constituents (Table 1.9), and hence, a residuum from an asphaltic (heavy, low-volatile, naphthenic) crude oil will be extremely complex.

Physical methods of selective fractionation usually indicate high proportions of nonvolatile **asphaltenes** and **resins** (Figure 1.14), even in amounts up to 50% (or higher) of the residuum. In addition, the presence of ash-forming metallic constituents, including such organometallic compounds as vanadium and nickel, is also a distinguishing feature of residua and the heavier oils. In the former case, the deeper the **cut** into the crude oil (i.e., the more volatile material that is removed), the greater is the concentration of sulfur and metals in the residuum, and the greater the deterioration in physical properties.

Asphaltenes (in particular) and resins (to a lesser extent) play a role in crude oil behavior in the reservoir and in refining. They can influence the efficiency of recovery operations, the

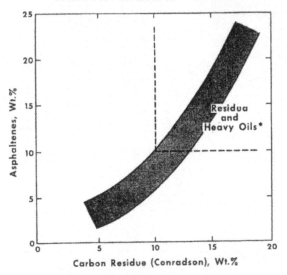

VARIATION OF THE YIELD OF THERMAL COKE (CARBON RESIDUE) WITH ASPHALTENE CONTENT

* Arbitrarily defined for the present purposes as having >10 wt. % asphaltenes and producing >10 wt. % thermal coke.

Figure 1.15: Fractionation of crude oil.

interaction of the crude oil with catalysts, as well as the amount of coke formed during processing (Figure 1.15).

   Asphalt is prepared from petroleum (Figure 1.16) and often physically resembles **bitumen** (often referred to as **native asphalt**). The process for asphalt manufacture essentially is a concentration of the nonvolatile materials in the crude oil into the nonvolatile residuum (Figure 1.17). Any changes are subject to the manner in which the asphalt is prepared from the residuum. Thus, there may be chemical similarities between the residuum and the asphalt, but differences must be anticipated. It is the chemical changes that can be the cause of instability or incompatibility of asphalt blends.

27

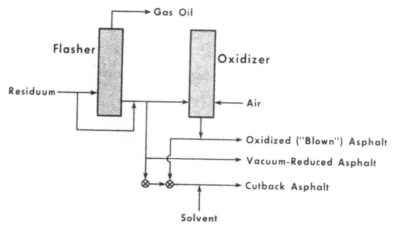

Figure 1.16: Asphalt manufacture.

It is recommended that the two types of asphalt be distinguished by use of the qualifying terms **native** and **petroleum**, except, in those occasional instances in which the source is other than petroleum, for example, **wurtzilite asphalt**.

When the asphalt is produced simply by distillation of petroleum or an asphaltic crude oil, the product can be referred to as **residual asphalt** or **straight-run asphalt**. If the asphalt is prepared by solvent extraction of residua or by light hydrocarbon (propane) precipitation, or if blown or otherwise treated, the term should be modified accordingly to qualify the product (e.g., **propane asphalt**).

By definition, any product that is produced by distillation without any further purification is referred to as the **straight-run** product.

Sour and **sweet** are terms referring to a crude oil's approximate sulfur content. In the past, these terms designated smell. A crude oil with a high sulfur content usually contains hydrogen sulfide, and the crude oil was called **sour**. Without this disagreeable odor, the crude oil was judged **sweet**.

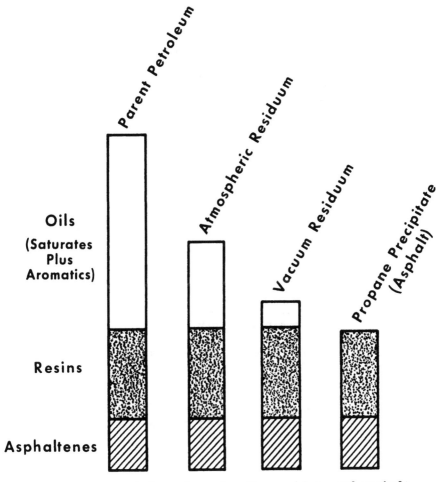

Figure 1.17: Composition of crude oil, residua, and asphalt.

Table 1.10 Classification of coal according to rank.

Class	Group	Limits of fixed carbon or Btu mineral-matter-free basis
I. Anthracite	Meta-anthracite	Dry FC >98% dry VM <2%
	Anthracite	Dry FC >92% but <98% dry VM <8% but >2%
	Semianthracite	Dry FC >80% but <92% dry VM <14% but <8%
II. Bituminous	Low-volatile bituminous coal	Dry FC >78% but <86% dry VM <22% but >14%
	Medium-volatile bituminous coal	Dry FC >69% but <78% dry VM <31% but >22%
	High-volatile A bituminous coal	Dry FC <69% dry VM >31%) moist Btu >14,000
	High-volatile B bituminous coal	Moist Btu >13,000 but <14,000
	High-volatile C bituminous coal	Moist Btu >10,000 but <13,000

III. Subbituminous	Subbituminous A coal	Moist Btu >11,000 but <13,000
	Subbituminous B coal	Moist Btu >9,500 but <11,000
	Subbituminous C coal	Moist Btu >8,300 but <9,500
IV. Lignite	Lignite	Moist Btu <8,300
	Brown coal	Moist Btu <8300

Source: 1993 Annual Book of ASTM Standards, American Society for Testing and Materials, 1916 Race Street, Philadelphia, PA 19103, Standard D388.

2.3 Coal Related Terms

The terminology/nomenclature that is in use at the present time is the result of many schemes that have been suggested over the years since coal came into prominence as the object of scientific study. In a more general sense, coal can be defined as an organic, rocklike natural product. The resemblance of coal to a rock is due to its physical nature and composition (Speight, 1994), and the inclusion of a "natural product" term in a general definition is not an attempt to describe coal as a collection of specific, and separate, natural product chemicals. Such chemical species are universally recognized and are more distinct chemical entities than coal.

The designation of coal as a natural product is no stretch of the truth and arises because of the oft-forgotten fact that coal is the result of the decay and maturation of floral remains (which are natural product chemicals) over geologic time. Indeed, the organic origins of coal and its organic composition are too often ignored (Speight, 1994). However, we next discuss more appropriate definitions of coal and the manner by which this complex natural product can be classified.

PETROLEUM PRODUCTS: INSTABILITY AND INCOMPATIBILITY

In brief, the relationship between the coalification process (Figure 1.18) and the different coal types has been recognized for some time. Thus, the need arose to describe each individual sample of coal in terms that would accurately (even adequately) depict the physical and/or chemical properties (Kreulen, 1948; van Krevelen and Schuyer, 1957; Francis, 1961). Consequently, the terminology that came to be applied to coal essentially came into being as part of a classification system, and it is difficult (if not impossible) to separate terminology from classification and to treat each as a separate subject. This is, of course, in direct contrast to the systems used for the nomenclature and terminology of petroleum, natural gas, and related materials (Speight, 1990, 1991), Indeed, coal classification stands apart in the field of fossil fuel science as an achievement that is second to none insofar as the system(s) allow classification of all the coals that are known on the basis of standardized parameters (Table 1.10).

Of particular importance here is the carbon content of the coal, which is part of the basis for the modern classification system of coal. The kinds of plant material from which the coal originated, the kinds of mineral inclusions, and the nature of the maturation conditions that prevailed during the metamorphosis of the plant material give rise to different **coal types**. Thus, whereas the carbon content of the world's coals varies over a wide range (ca. 75.0-95.0% w/w), petroleum, on the other hand, does not "enjoy" (perhaps, fortunately) such a wide variation in carbon content. All of the **petroleum, heavy oil, and bitumen (natural asphalt)** that occur throughout the world fall into the range of 82.0-88.0% w/w. Hence, little room is left for the design of a standardized system of petroleum classification and/or nomenclature based on carbon content.

It is necessary to note that this particular aspect of coal science (petrography) deals with the individual components of coal as an organic rock, whereas the nomenclature, terminology, and classification systems are intended for application to the whole coal. Other general terms that are often applied to coal include **type, rank,** and **grade,** which are three terms that describe the particular characteristics of coal. The **rank** of a coal refers to the degree of metamorphosis. For example, coal that has undergone the most extensive change, or metamorphosis, has the highest rank (determined from the fixed carbon or heating value). The **grade** of a coal refers to the amount and kind of inorganic material (mineral matter) within the coal matrix. Sulfur is, perhaps, the most significant of the nonhydrocarbon materials because of the potential to generate sulfur dioxide during combustion.

Coal, in the simplest sense, consists of vestiges of various organic compounds that were originally derived from ancient plants and have subsequently undergone changes in molecular and physical structures during the transition to coal.

There have been attempts to classify coal as one of the hydrocarbon resources of the earth (Figure 1.3); the term **hydrocarbon** is used very generally since atoms (nitrogen, oxygen, and sulfur) other than carbon and hydrogen are present in these materials. Furthermore, the classification of petroleum into various subgroups (Table 1.3) and the classification of petroleum, for example, by use of density-gravity data (Table 1.4) offer some hope for the design of a classification system for petroleum.

One general system of nomenclature is a system by which coal can be divided into four major types: **anthracite coal**, **bituminous coal**, **subbituminous coal**, and **lignite**. Peat is usually not classified as a coal and, therefore, is not included in this system (Figure 1.18).

Anthracite coal is coal of the highest metamorphic rank; it is also known as "hard" coal and has a brilliant luster. It can be rubbed without leaving a familiar "coal dust" mark on the finger and can even be polished for use as jewelry. Anthracite coal burns slowly with a pale blue flame and may be used primarily as a domestic fuel.

On the other hand, bituminous coal burns with a smoky flame and may also contain 15-20% w/w volatile matter. It is the most abundant variety of coal, weathers only slightly, and may be kept in open piles with very little danger of spontaneous combustion. However, there is evidence that spontaneous combustion is more a factor of extrinsic conditions such as the mining and storage practices and the prevalent atmospheric conditions.

Subbituminous coal is not as high on the metamorphic scale as bituminous coal and has often been called **black lignite**.

Finally, lignite is the coal that is lowest on the metamorphic scale. It may vary in color from brown to brown-black and is often considered to be intermediate between peat and the subbituminous coal. Lignite is often distinguished from subbituminous coal by

	Composition, wt % (DAF Basis)		
	C	H	O
Wood	49	7	44
Peat	60	6	34
Lignite	70	5	25
Subbituminous	75	5	20
Bituminous	85	5	10
Anthracite	94	3	3

Increasing Aromatization;
Loss of Oxygen

⟶

Increasing Pressure/
Temperature/Time

Figure 1.18: The coalification process.

having a lower carbon content and a higher moisture content. Lignite may dry out and crumble in air and is certainly liable to spontaneous combustion.

Thus, it is obvious that this general system of nomenclature offers little, if anything, in the way of a finite description of the various coals. In fact, to anyone but an expert (who must be presumed to be well versed in the field of coal technology), it would be extremely difficult to distinguish one piece of coal from another. Therefore, the terminology that is applied to coal is much more logical when it is taken in perspective with the classification system from which it has arisen and becomes much more meaningful in terms of allowing specific definitions of the various types of coal that are known to exist.

Thus, there is a need to accurately describe the various coals in order to identify the end use of the coal and also to provide data that can be used as a means of comparison of the various worldwide coals. Hence, it is not surprising that a great many methods of coal classification have arisen over the last century or so. These methods were, of course, designed to serve many special interests, ranging from the more academic classification systems of geologists, paleobotanists, and chemists to the more commercial types of classification used by utility company personnel. However, it became essential as coal rose to a prominent position as a fuel that a system be accepted which would be applied (almost) universally and the terminology of which would allow immediate recognition of a coal type, irrespective of the place or country of origin of the coal.

The widespread and diverse uses of coal have, as stated above, resulted in the development of numerous classification systems. Indeed, these systems have invoked the use of practically every chemical and physical characteristic of coal. Consequently, the literature available for those wishing to make a thorough study of the various classification systems is voluminous, and it would be futile here to enter upon a discussion of all those systems.

Nevertheless, it is necessary to describe the major classification systems in current use. Several of the lesser known classification systems are also included because, although they may not be recognized as part of a more formalized classification system, they often contain elements of coal terminology that may be in current use by the various scientific disciplines involved in coal technology.

Coal contains significant proportions of carbon, hydrogen, and oxygen, with lesser amounts of nitrogen and sulfur. Thus, it is not surprising that several attempts have been made to classify coal on the basis of elemental composition. Indeed, one of the earlier classifications of coal, based on its elemental composition, was subsequently extended. This system (Speight, 1994) offered a means of relating coal composition to technological properties and may be looked upon as a major effort to relate properties to utilization. Indeed, for coal below the anthracite rank, and with an oxygen content less than 15%, it was possible to derive relationships between carbon content ($C\%$ w/w), hydrogen content ($H\%$ w/w), calorific value (Q, cal g^{-1}), and volatile matter ($VM\%$ w/w):

$$C = 0.59(Q/100 - 0.367VM) + 43.4$$
$$H = 0.069(Q/100 + VM) - 2.86$$
$$Q = 388.12H + 123.92C - 4269$$
$$VM = 10.61H - 1.24C + 84.15$$

Since these relationships only apply to specific coal types their application is limited, and it is unfortunate that composition and coal behavior do not exist in the form of simple relationships. In fact, classification by means of elemental composition alone is extremely difficult.

Nevertheless, this attempt at coal classification is of interest in the present context insofar as the system contained a term for volatile matter production, which can also be equated to liquids production. Some allowance must be made for gas production, but it is possible to estimate the character of the liquids knowing the analytical data for the parent coal, the yield and character of the gases, and the yield and character of the coke.

Furthermore, it is worthy of note that coal rank is often equated directly with carbon content. Even though this may actually appear to be the case, classification of coal by rank is a progression from high-carbon coal to low-carbon coal (or vice versa), but this is in conjunction with other properties of the coal.

The American Society for Testing and Materials (ASTM) (1993) has evolved a method of coal classification (ASTM D 388) over the years that relied on the fixed carbon content of coal as well as other physical properties (Table 1.10). The **fixed carbon** is the solid residue, other than mineral ash, remaining after all of the **volatile matter** has been removed under prescribed conditions. It is often simply described as a coke-like residue. The value is calculated by subtracting moisture, volatile matter, and ash from 100% carbon content.

The basis for such a classification is the proximate analysis of the coal, resulting in the division of all coals into a series of classes and groups. Coal with a rank higher than high-volatile B bituminous coal (class II, group 4), i.e., coal having less than 31% w/w volatile matter, is classified according to the fixed carbon content; coal having more than 31% w/w volatile matter (daf) is classified according to the moist calorific value. Moist calorific value is the calorific value of the coal when the coal contains its natural bed moisture. The natural bed moisture is often determined as the equilibrium moisture under a set of standard conditions. In addition, the agglomerating characteristics of coal are used to differentiate between certain adjacent groups. This system of classification, in fact, indicates the degree of coalification as determined by these methods of proximate analysis, with lignite being classed as low-rank coal; the converse applies to anthracite. Thus, coal rank increases with the amount of fixed carbon but decreases with the amount of moisture and volatile matter. It is, perhaps, easy to understand why coal rank is often (and incorrectly) equated to changes in the proportion of elemental carbon in coals (ultimate analysis).

It is true, of course, that anthracite coal usually contains more carbon than bituminous coal, which in turn, usually contains more carbon than subbituminous coal, and so on. Nevertheless, the distinctions between the proportions of elemental carbon in the various coals are not as well defined as for the fixed carbon, and extreme caution is advised in attempting to equate coal rank with the proportion of elemental carbon in the coal.

Finally, because the petroleum industry recognizes that there are many fractions analogous to **tar** and its derivatives, it is worth considering the definitions here in more detail. In more general terms, **tar** is the result of the destructive distillation of many bituminous or other organic materials and is a brown to black, oily, viscous liquid. Tar is most commonly produced from bituminous coal and is generally understood to refer to the coal product, although it is advisable to specify coal tar if there is the possibility of ambiguity.

Coal tar is often the source of coal liquids, and many of the liquids produced from coal contain constituents that are molecularly similar (homologous) to those present in the coal tar.

The most important factor in determining the yield and character of the coal tar is the carbonizing temperature. Three general temperature ranges are recognized, and the products have acquired the designations low-temperature tar (ca. 450-700°C; 840-1290°F); midtemperature tar (ca. 700-900°C; 1290-1650°F); and high-temperature tar (ca. 900-1290°C; 1650-2190°F).

Tar released during the early stages of the decomposition of the organic material is called primary tar, since it represents a product that has been recovered without the secondary alteration that results from prolonged residence of the vapor in the heated zone.

Treatment of the distillate (boiling up to 250°C, 480°F) of the tar with caustic soda causes separation of a fraction known as tar acids; acid treatment of the distillate produces a variety of organic nitrogen compound known as tar bases. The residue left following removal of the heavy oil, or distillate, is pitch, a black, hard, and highly ductile material.

2.4 Oil Shale Related Terms

Oil shale comprises an enormous and largely untapped fossil fuel resource. As readily accessible petroleum sources dwindle because of economic or political issues, utilization of the oil shale resource to meet world needs for energy and chemical feedstocks will become both necessary and economically attractive.

Worldwide oil shale deposits are estimated to contain 30 trillion barrels of shale oil, but only a small fraction of this amount is easily recoverable using current technology.

Thus, the utilization of oil shale to replace petroleum will mean finding economically efficient and environmentally acceptable methods for recovering the energy-rich organic material locked inside the oil shale's rock matrix and for upgrading the recovered shale oil (Scouten, 1990).

Oil shale is a complex and intimate mixture of organic and

inorganic materials that varies widely in composition and properties. Some, such as the **oil shale** of the Green River Formation, are not even true shales.

Oil shale technology has a long history. Oil shale was a source of oil as early as AD 800 and the British oil shale deposits were worked in Phoenician times. The use of oil shale was recorded in Austria in 1350. The first shale oil patent (British Crown Patent 330) was issued in 1694 to Martin Eele, Thomas Hancock, and William Portlock, who "after much paines and expences hath certainely found out a way to extract and make great quantityes of pitch, tarr and oyle out of a sort of rock."

Oil shale utilization on an industrial scale did not, however, follow immediately. Not until 1838 was the first industrial oil shale plant put into service at Autun, France. Plants soon followed in Scotland (1850), Australia (1865), and Brazil (1881). In addition to being a source of refined shale oil products, it was soon discovered that **torbanite**, an especially organic-rich type of oil shale, was useful for increasing the luminosity of gas flames. This provided an important market for the early Scottish oil shale industry and later, as the Scottish torbanite deposits were depleted, for the early Australian shale industry.

By the 1870s, Australian torbanite was being exported not only to Great Britain, but also to the United States, Italy, France, and Holland. The invention of the Welsbach gas mantle and the advent of low-cost, high-quality kerosene from American petroleum spelled the end of this period. As the need for liquid transportation fuels increased, the Australian oil shale operations consolidated. Elsewhere, oil shale plants followed in New Zealand (1900), Switzerland (1915), Sweden (1921), Estonia (now USSR, 1921), Spain (1922), China (1929), and South Africa (1935).

The high point of this stage of oil shale development was reached during, or just after, World War II. However, the oil shale industry in Estonia and neighboring Leningrad Province still flourishes, most of the mined shale being burned directly in electric power generating plants and the remaining 10% or so retorted to provide chemical feedstock and smaller quantities of refined products.

In 1926 the Japanese began commercial production of shale oil from the large Chinese oil shale deposits at Fushun in Manchuria. Improved retorts were installed at this complex in 1941 to provide important supplies of liquid fuels for the Japanese forces during World War II. At Maoming, near Canton in southern China, a second oil shale project was developed. Shale oil production in the People's Republic of China peaked about 1975 and has since declined, as emphasis has shifted to newly discovered petroleum supplies.

The near-term future of oil shale is uncertain. Very clearly, this future will be influenced by international crude oil prices and supplies. Indeed, the rise of interest in oil shale during the late 1970s was due largely to the high prices and tight supply of crude oil. With the decline of crude prices and the politics behind the development of petroleum resources and the pricing of petroleum (Yergin, 1992), interest in oil shale and other synfuels waned. How long this will continue is difficult to predict. However, it is obvious that "foreign" petroleum will be priced at a level that makes the development of domestic synthetic fuels uneconomical.

At present, commercial oil shale development is hard to foresee unless impetus is provided by political or security concerns. However, the current lull in development activity offers a golden opportunity for scientific research to attack the many chemical, physical and material problems that were uncovered or brought into sharper focus during the late period of activity.

Politics aside, there is no scientific definition of **oil shale**; the definition is strictly an economic one: oil shale is a compact, laminated rock of sedimentary origin, yielding over 33% ash and containing organic matter that yields oil when distilled but not appreciably when extracted with the ordinary solvents for petroleum.

Simpler definitions include use of the term **oil shale** to denote an organic-rich rock that contains little or no free oil, and the term **shale oil** is defined as the oil produced from an oil shale on heating.

Three other terms will used extensively; hence, their definitions are important: **bitumen** (remembering above, the use of the term bitumen to describe a **natural asphalt**) is defined as the organic material that can be extracted by ordinary organic solvents, such as benzene (C_6H_6), toluene ($C_6H_5 \cdot CH_3$), tetrahydrofuran (C_4H_8O), and chloroform (C_3HCl), or mixtures (generally an azeotrope) of such solvents, such as benzene-methanol (60:40). In fact, the term **bitumen** as used for the extract from virgin oil shale may be quite correct insofar as it is used in the same sense as when it is applied as an alternate term for natural asphalt. However, when applied to the corresponding fraction from a retorted oil shale, a term analogous to **pitch** would be more correct. Differentiation through the use of qualifying terms such as **coal tar pitch** and **oil shale pitch** (for material separated from retorted oil shale) would be more appropriate.

Kerogen, which makes up the major part of the organic material, is not soluble in such solvents. Kerogen is the name given to what is purportedly the precursor to petroleum, although such a definition is open to speculation and criticism (Speight, 1991). Kerogen is an organic material found in shale deposits, but whether or not it is a true precursor to oil has not been proven.

Table 1.11   Analytical data for various kerogen concentrates

wt%	Green River (USA)	Alexsinac (Yugoslavia)	Irati (Brazil)	Pumpherston (Scotland)
Carbon	77.4	71.9	78.8	75.7
Hydrogen	10.3	8.7	9.5	10.4
Nitrogen	3.1	3.2	3.9	4.2
Oxygen + sulfur	9.2	16.2	7.8	9.7

It is this type of terminology that adds more confusion to an already confusing system of nomenclature because of the presumed interrelationships between various types of kerogen and petroleum (Figure 1.19). For example, **kerogen** is a precursor to a **bitumen,** which again, is a case of misnomenclature, or mistaken identity. It is preferable that the term **bitumen** be restricted in use to the naturally occurring material that is closely related to petroleum. "Soluble kerogen" would be a better term for the "bitumen" derived from kerogen, as it would serve to qualify the origin of the material.

Kerogen can be isolated from the inorganic matrix as a **kerogen concentrate** (Table 1.11), which refers to the organic concentrate that is produced by beneficiation or chemical demineralization of an oil shale. Strictly speaking, this term should refer only to that part of the organic concentrate that is insoluble in organic solvents. However, in common usage, kerogen concentrate refers to the total organic material (kerogen + bitumen) obtained by removing minerals.

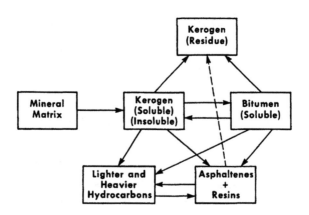

Figure 1.19: Kerogen-crude oil relationships.

These are operational definitions, and the relative proportions of bitumen and kerogen depend on the choice of solvent and extraction conditions. Nevertheless, these are useful definitions, provided their limitations are kept in mind.

Such definitions must be used with some caution, since they can only provide so much in terms of understanding a scientific and engineering concept or phenomenon. Nevertheless, they do serve a purpose when used in the correct context.

3.0 References

American Society for Testing and Materials 1993 ASTM D 388. Classification of Coal by Rank. 1993 Annual Book of ASTM Standards. Section 05.05., Philadelphia, Pennsylvania.

Barth, E.J. 1962. Asphalt: Science and Technology. Gordon and Breach, New York.

Brinkman, D.W., and White, E.W. 1981. In Distillate Fuel Stability and Cleanliness. L.L. Stavinoha and C.P. Henry (editors). Special Technical Publication No. 751. American Society for Testing and Materials, Philadelphia, Pennsylvania. P. 84.

Broome, D.C., and Wadelinn, F.A. 1973. In Criteria for Quality of Petroleum Products. J.P. Allinson (editor). Halsted Press, Toronto. Chapter 13.

Evans, C.R., Rogers, M.A., and Bailey, N.J.L. 1971. Chem. Geol. 8: 147.

Francis, W. 1961. Coal: Its Formation and Composition. Edward Arnold Ltd., London.

Gray, M.R. 1994. Upgrading Petroleum Residues and Heavy Oils. Marcel Dekker Inc., New York.

Hardy, D.R., and Wechter, M.A. 1990. Fuel. 69: 720.

Hoiberg, A.J. (editor). 1964. Bituminous Materials: Asphalts, tar, and Pitches. Volumes I, II, and III. Interscience Publishers, New York.

Kreulen, D.J.W. 1948. Elements of Coal Chemistry. Nijgh and van Ditmar N.V., Rotterdam, The Netherlands.

McKetta, J.J. (Editor) 1992. Petroleum Processing Handbook. Marcel Dekker Inc., New York.

Mochida, J., Sakanishi, K., and Fujitsu, H. 1986. Oil Gas. November.

National Research Council. 1990. Fuels to Drive Our Future. National Academy Press, Washington, DC.

Por, N. 1992. Stability Properties of Petroleum Products. Israel Institute of Petroleum and Energy, Tel Aviv, Israel.

Power, A.J., and Mathys, G.I. 1992. Fuel. 71: 903.

Ruzicka, D.J., and Nordenson, S. 1990. Fuel. 69: 710.

Scouten, 1990. In Fuel Science and Technology Handbook. J.G. Speight (editor). Marcel Dekker Inc., New York. Chapter 25.

Speight, J.G., (editor). 1990. Fuel Science and Technology Handbook. Marcel Dekker Inc., New York.

Speight, J.G. 1991. The Chemistry and Technology of Petroleum. 2nd Edition. Marcel Dekker Inc., New York.

Speight, J.G.. 1994. The Chemistry and Technology of Coal. 2nd Edition. Marcel Dekker Inc., New York.

Stavinoha, L.L., and Henry, C.P. (editors). 1981. Distillate Fuel Stability and Cleanliness. Special Technical Publication No. 751. American Society for Testing and Materials, Philadelphia, Pennsylvania.

Traxler, R.N. 1961. Asphalt: Its Composition, Properties, and Uses. Reinhold Publishing Corp., New York.

van Krevelen, D.W., and Schuyer, J. 1957. Coal Science: Aspects of Coal Constitution. Elsevier, Amsterdam.

Wallace, T.J. 1964. In Advances in Petroleum Chemistry and Refining. J.J. McKetta Jr. (editor). Interscience, New York. P. 353.

Yergin, D. 1992. The Prize. Simon & Schuster, New York.

CHAPTER 2: SOURCES OF LIQUID FUELS. I. PETROLEUM

1.0 Introduction

The best means of grasping the issue of incompatibility is from an understanding of the nature of the sources of liquid fuels. Furthermore, understanding the means by which liquid fuels and other products are produced as well as the composition of the products provides insights into the chemistry and physics of instability and incompatibility.

Petroleum (Gruse and Stevens, 1960; Hobson and Pohl, 1973; Speight, 1991; McKetta, 1992) is the predominant source of liquid fuels; coal (Speight, 1994) and oil shale (Scouten, 1990) have also provided liquid fuels on demand, perhaps not economically. Indeed, inspections of various shale oils indicate that high nitrogen contents are the norm (relative to petroleum) (Table 2.1) (Scouten, 1990).

In contrast, liquid fuels from coal contain substantial

Table 2.1: Approximate composition of unrefined shale oil.

Property	Range
Gravity, °API	10-25
Pour point, °F	-11-85
Elemental analysis, wt%	
carbon	83-85
hydrogen	10-12
nitrogen	1- 2
oxygen	1- 2
sulfur	1- 5
Metals, ppm	
vanadium	0-150
nickel	2- 10
arsenic	30-100
iron	45-110
Asphaltenes, wt%	0- 1
Boiling range (>95% recovery)	
°C	32-530
°F	90-985

amounts of oxygen, usually in the form of phenols. Therefore, both alternate sources of liquid fuels (oil shale and coal) offer the potential for increased costs to remove the offending heteroatom species. If this is not done, instability and incompatibility will result. And this is not even considering any environmental issues that arise from the presence of such species in fuels.

It is, however, not only a matter of cost to remove such species, but it is a matter for the various levels of government to decide the price they are willing to pay for a level of security of supply of fuels from domestic sources.

Thus, it is the potential for the use of coal and oil shale that dictates some study in this text. Should they ever be employed as substantial sources of fuels and other products, the question of compatibility with similar (boiling range) products from petroleum will arise.

Indeed, this question did arise during the late 1970s and early 1980s, when emphasis was placed on coal and oil shale as alternate feedstocks to petroleum. The nature of the refinery operations and the technological adaptations required to handle such products were of some concern.

The concern was based on the reasonable assumption that any product streams from coal refining and from oil shale refining would be further treated in a petroleum refinery. The need to dedicate specific refineries to coal and oil shale was seen as an expense that would only add to the ultimate cost of the products to the consumer.

Thus, it is not only necessary to consider the variations in product composition when different types of petroleum are employed, but there is also the need to consider coal (Chapter 3) and oil shale (Chapter 4) as sources of fuels and other products where the issues of instability and incompatibility could well arise.

Petroleum is a mixture of gaseous, liquid, and solid hydrocarbon-type chemical compounds that occur in sedimentary rock deposits throughout the world.

In the crude state, petroleum has little value, but when refined, it provides liquid fuels, solvents, lubricants, and many other products using a variety of processes with varying chemistry (Chapter 1). The fuels derived from petroleum contribute a substantial portion (30-50%) of the total (world) supply of energy and are used not only for transportation fuel (gasoline, diesel

Table 2.2: Crude petroleum is a mixture of compounds that can be
separated into different generic boiling fractions.

Fraction	Boiling Range	
	°C	°F
Light naphtha	-1-150	30-300
Gasoline	-1-180	30-355
Heavy naphtha	150-205	300-400
Kerosene	205-260	400-500
Stove oil	205-290	400-550
Light gas oil	260-315	400-600
Heavy gas oil	315-425	600-800
Lubricating oil	>400	>750
Vacuum gas oil	425-600	800-1050
Residuum	>600	>1050

fuel, aviation fuel, etc.), but also for domestic and commercial
heating (fuel oil). Beyond this point, products from petroleum are
also employed for use as domestic and commercial lubricants as well
as the more familiar highway and roofing asphalt. These latter
products are often less recognized as being susceptible to
instability and incompatibility issues. Nevertheless, such issues
do arise (Chapter 5).

Crude petroleum is a mixture of several series of compound
types that boil over a wide range of temperatures (Figure 2.1) and
that can be separated into a variety of different generic fractions
(Table 2.2). Since there is a wide variation in the properties of
crude petroleum (Speight, 1991), the proportions in which the
different fractions occur and the molecular species within the
fractions can vary with origin. Thus, some crude oils have higher
proportions of the lower-boiling components, whilst others have
higher proportions of residuum (asphaltic components). Molecular
variations in the character of the constituents complicate the
issue still further.

The outcome of petroleum refining is the production of a heavy
(nondistillable) product and a light (distillable) product (Figure
2.2). The generation of light products that are not indigenous to

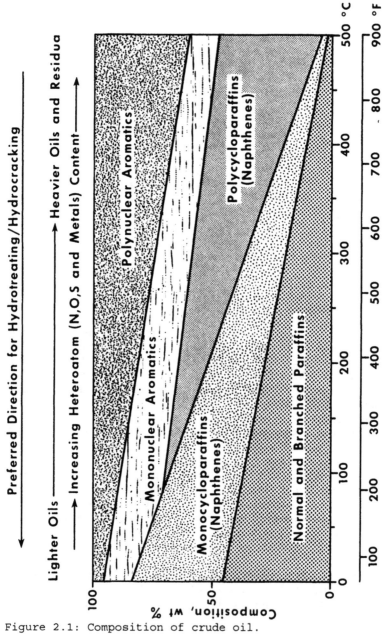

Figure 2.1: Composition of crude oil.

48

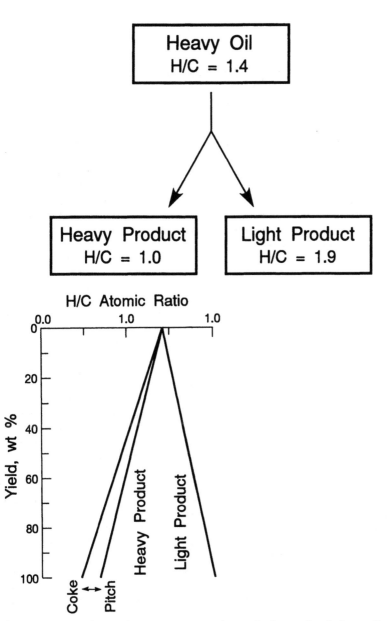

Figure 2.2: Schematic representation of the principles of petroleum refining.

the crude oil can be achieved by the intramolecular movement of hydrogen or by the addition of hydrogen, as in the case of the various hydroprocesses (see below).

It is generally assumed, but it is not always correct, that the high-molecular-weight and high-boiling material give rise to coke, whilst the low-molecular-weight and low-boiling material give rise to distillate products (Figure 2.3). However, this is not always true. For example, asphaltenes can give rise to distillable products to the extent of ca. 50% w/w yield, and lower-boiling materials, such as gas oil, can yield coke. The use of indigenous hydrogen by intramolecular reactions promotes distillate formation.

The intramolecular movement of hydrogen to produce a hydrogen-rich product (distillate) and a hydrogen-poor product (coke or pitch) is, in itself, an interesting concept and, possibly, a prelude to incompatibility. The generation of the highly aliphatic material and the highly aromatic material may be sufficient reason for incompatibility to occur in many thermal processes.

Petroleum refining often involves blending crude oils from several wells and thereby homogenizing the feedstock to the refinery. It is usual practice to blend crude oils of similar characteristics, although fluctuations in the properties of the individual crude oils may cause significant variations in the properties of the blend over a period of time.

The concept of crude oil blending is considered to simplify refining since the refinery can accept an "average" feedstock more economically than the operations (processes) can be "adjusted" to accept individual feedstocks on an irregular basis. Blending several crude oils may occur prior to transportation, or even after transportation but prior to refining.

At this point, it is worthy of note that one aspect of petroleum occurrence and resource development that is often ignored relates to the presence of natural gas.

Natural gas is a combustible gas that occurs in porous rock in the earth's crust and is found with, or near, crude oil reservoirs. However, it may occur alone in separate reservoirs. More commonly, it forms a gas cap (cap rock) entrapped between liquid petroleum and an impervious rock layer (cap rock) in a petroleum reservoir. Under conditions of greater pressure, the gas will be intimately mixed with, or dissolved in, crude oil.

Typical natural gas consists of hydrocarbons having a very low

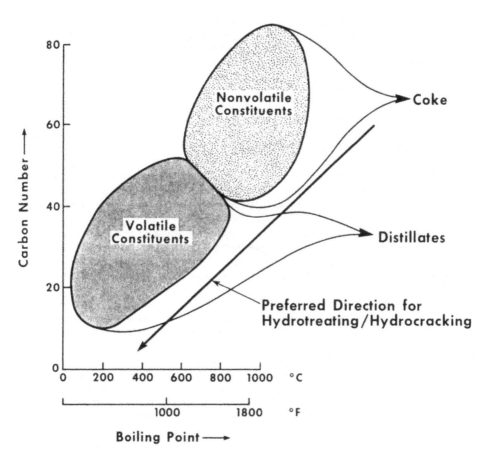

Figure 2.3: Comversion of crude oil to distillate and coke.

boiling point. Methane (CH_4), the first member of the paraffin series, and with a boiling point of -159°C (-254°F), makes up approximately 85% of the gas. Ethane (C_2H_6), with a boiling point of -89°C (-128°F), may be present in amounts up to 10% and propane, (C_3H_8), with a boiling point of -42°C (-44°F) may also be present. Dissolution of the constituents of natural gas in petroleum under pressure can cause instability or incompatibility in the reservoir, manifested by the deposition of asphaltic (asphaltene and resin) constituents (Evans et al., 1971). The process is analogous to the deasphalting process, in which adding a low-boiling liquid hydrocarbon to petroleum will cause separation of the asphaltic constituents (Speight, 1991).

Petroleum refining is the separation of petroleum into fractions and the subsequent treating of these fractions to yield marketable products. Most petroleum products are either fractions of petroleum that have been treated to remove undesirable components or are totally, or partly, synthetic, in that they have compositions that are impossible to achieve by direct separation of these materials from crude petroleum (Nelson, 1958; Bland and Davidson, 1967; Speight, 1991). They result from chemical processes that change the molecular nature of selected portions of crude petroleum.

For example, the manufacture of products from the lower-boiling portion of petroleum automatically produces a certain amount of higher-boiling components. If the latter cannot be sold as, say, heavy fuel oil, these products will accumulate until refinery storage facilities are full.

To prevent the occurrence of such a situation, the refinery must be flexible and be able to change operations as needed. This usually means more processes: thermal processes to change an excess of heavy fuel oil into more gasoline (with coke as the residual nonvolatile product) or a vacuum distillation process to separate the heavy oil into lubricating oil stocks and asphalt.

The outcome is, in the case of the thermal processes, the change in the molecular structure of many of the constituents. Thus, although petroleum refining is a delicately balanced system where the constituents coexist in an equilibrium state, any changes can cause instability of the system, and the outcome can be the production of an incompatible fraction either during refining or after product separation.

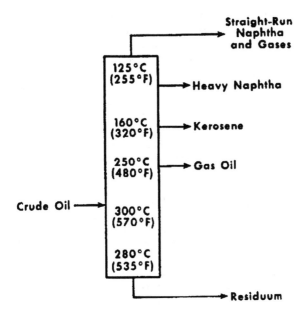

Figure 2.4: Distillation of petroleum.

The goal of this chapter is to review the methods by which liquid fuels and other products are manufactured from petroleum. This will give an understanding of the background of petroleum refining as well as an indication of the chemistry of conversion. From such an understanding, the various compounds in the products can be identified and predictions about instability and incompatibility can be made.

2.0 Distillation

Distillation was the first method by which petroleum was refined. The evolution of the distillation process has played a major role in the development of the refining industry (Speight, 1991).

It is possible to obtain products ranging from gaseous materials taken off at the top of the distillation column to a nonvolatile residue or reduced crude ("bottoms"), with correspondingly lighter materials at intermediate points (Figure 2.4). The reduced crude (atmospheric residuum) may then be processed by vacuum, or steam, distillation in order to separate the high-boiling lubricating oil fractions without the danger of decomposition, which occurs at high (>350°C, >660°F) temperatures.

Atmospheric distillation may be terminated with a lower-boiling fraction ("cut") if it is felt that vacuum or steam distillation will yield a better-quality product, or if the process appears to be economically more favorable. Not all crude oils yield the same distillation products, and the nature of the crude oil dictates the processes that may be required for refining.

2.1 Atmospheric Distillation

The present-day petroleum distillation unit is, in reality, a collection of distillation units, but in contrast to the early battery units, a tower is used in the modern-day refinery and brings about a fairly efficient degree of fractionation (separation).

The feed to a distillation tower is heated by flow through pipes arranged within a large furnace. The heating unit is known as a pipe still heater or pipe still furnace, and the heating unit and the fractional distillation tower make up the essential parts of a distillation unit or pipe still. The pipe still furnace heats the feed to a predetermined temperature, usually a temperature at which a predetermined portion of the feed will change into vapor. The vapor is held under pressure in the pipe in the furnace until it discharges as a foaming stream into the fractional distillation tower. Here the unvaporized or liquid portion of the feed descends to the bottom of the tower to be pumped away as a bottom nonvolatile product, while the vapors pass up the tower to be fractionated into gas oil, kerosene, and naphtha.

Conversion of the petroleum to thermal products is not a characteristic of the pipe still operations, although some changes can occur in the functional group distribution (Speight and Francisco, 1990). It should be possible to blend the components back together (in the correct proportions) and form the original crude oil without any sign of instability.

All of the primary fractions from a distillation unit are equilibrium mixtures and contain some proportion of the lighter constituents characteristic of a lower-boiling fraction. The primary fractions are "stripped" of these constituents ("stabilized") before storage or further processing.

2.2 Vacuum Distillation

Vacuum distillation evolved because of the need to separate the less volatile products, such as lubricating oils, from the petroleum without subjecting these high-boiling products to cracking conditions. The boiling point (b.p.) of the highest boiling fraction obtainable at atmospheric pressure is limited by the temperature (ca. 350°C; ca. 660°F) at which the residue starts to decompose or "crack." Above this temperature, the rate of cracking increases phenomenally (Chapter 1), and products are formed that often bear no chemical or physical relationship to the original constituents of the crude oil. This results in the oft observed occurrence of immiscibility (incompatibility) or instability (due to oxidation) of the thermal products.

Operating conditions for vacuum distillation are usually at a pressure of the order of 50-100 mm of mercury (atmospheric pressure = 760 mm of mercury = 14.8 psi). In order to minimize large fluctuations in pressure in the vacuum tower, the units are necessarily of a larger diameter than the atmospheric units. Some vacuum distillation units have diameters of the order of 45 feet (14 meters). By this means, a heavy gas oil may be obtained as an overhead product at temperatures of about 150°C (300°F), and lubricating oil cuts may be obtained at temperatures of 250-350°C (480-660°F), feed and residue temperatures being kept below 350°C (660°F), above which cracking will occur. The partial pressure of the hydrocarbons is effectively reduced still further by the injection of steam. The steam added to the column, principally for the stripping of asphalt in the base of the column, is superheated in the convection section of the heater.

By the use of various distillation pressures, a series of residua (Table 2.3) can be prepared, depending upon the needs of the refinery and the next-step processing of the residua. However, in general, the production of residua focuses on an atmospheric residuum (b.p. >345°C; >650°F) and a vacuum residuum (b.p. >510°C; >950°F or b.p. >565°C; >1050°F). Unless a real need, or market, develops, there is a hesitancy by refiners to move away from these boiling ranges for residua.

Table 2.3: Properties of residua (Tia Juana light crude oil)

Property								
Boiling range	°F	whole	>430	>650	>750	>850	>950	>1050
	°C	crude	>220	>345	>400	>455	>510	> 565
Yield		100	70	49	40	31	24	18
Gravity, °API		32	23	17	16	13	10	7
Nitrogen				0.4			0.5	0.6
Sulfur		1.0				2.1	2.4	2.6
Carbon residue								
Conradson			7	9	11	14	17	22

Finally, there is no guarantee that two similar cut-point residua from different crude oils will be chemically and physically similar. In fact, differences might be anticipated (Figure 2.5) based on local and regional differences in the precursors that formed the crude oils as well as local and regional differences in the maturation process.

Such differences are manifested in the form of different processing requirements and might also be the preliminary aspects of incompatibility.

2.3 Azeotropic and Extractive Distillation

As refineries evolved, distillation techniques became more sophisticated to handle a wider variety of crude oils to produce marketable products or feedstocks for other refinery units.

However, it became apparent that the distillation units in the refineries were incapable of producing specific product fractions. In order to accommodate this type of product demand, refineries have, in the latter half of this century, incorporated azeotropic distillation (Figure 2.6) and extractive distillation (Figure 2.7) in their operations.

All compounds have definite boiling temperatures, but a mixture of chemically dissimilar compounds will sometimes cause one or both of the components to boil at a temperature other than that expected. A mixture that boils at a temperature lower than the boiling point of any of the components is an azeotropic mixture.

When it is desired to separate close-boiling components, the

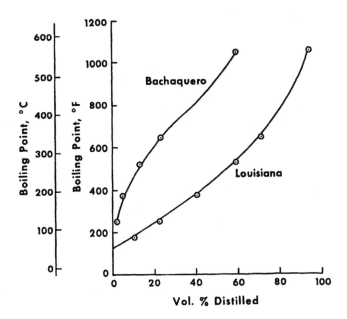

Properties of the 1050°F + Residua:

	Louisiana	Bachaquero
Gravity, API	12.1	2.8
Sulfur, wt. %	0.50	3.71
Nitrogen, wt. %	0.42	0.60
MNI, wt. %	10.5	20.0
Con. Carbon, wt. %	15.8	27.5
Nickel, ppm	26	100
Vanadium, ppm	19	900
Pour Point, °F	-	130

Figure 2.5: Distillation profiles for two crude oils.

addition of a nonindigenous component will form an azeotropic mixture with one of the components of the mixture, thereby lowering the boiling point by the formation of an azeotrope and facilitating separation by distillation.

The separation of these components of similar volatility may become economic if an "entrainer" can be found that effectively changes the relative volatility. It is also desirable that the entrainer be reasonably cheap, stable, nontoxic, and readily recoverable from the components. In practice, it is probably this last-named criterion that limits severely the application of extractive and azeotropic distillation.

The majority of successful processes are those in which the entrainer and one of the components separate into two liquid phases on cooling if direct recovery by distillation is not feasible. A further restriction in the selection of an azeotropic entrainer is that the boiling point of the entrainer be in the range of 10 to 40°C (18 to 72°F) below that of the components.

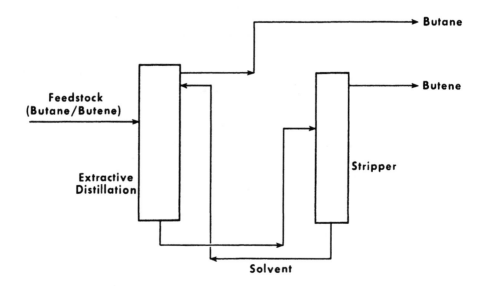

Figure 2.6: Extractive distillation.

3.0 Cracking Processes

Cracking was used commercially in the production of oils from coal and shales before the petroleum industry began, and the discovery that the heavier products could be decomposed into lighter oils, a process called cracking distillation, was used to increase the production of kerosene.

The yields of kerosene were usually markedly increased by means of cracking distillation, but the technique was not suitable for gasoline production. As the need for gasoline arose in the early 1900s, the necessity of prolonging the cracking process became apparent and a process known as pressure cracking evolved.

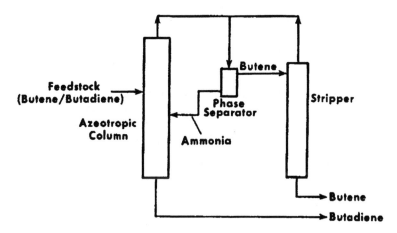

Figure 2.7: Azeotropic distillation.

Pressure cracking was a batch operation in which, as an example, gas oil (200 billion barrels (bbl)) was heated to about 425°C (800°F) in stills that had been reinforced to operate at pressures as high as 95 psi (6.4 atmospheres). The gas oil was held under maximum pressure for 24 hr., while fires maintained the temperature. Distillation was then started, and during the next 48 hr. produced a lighter distillate (100 bbl), which contained the gasoline components. However, instability was recognized even at this time, and the distillate was treated with sulfuric acid to remove unstable gum-forming components and then redistilled to produce a cracked gasoline.

Further developments in, and evolution of, cracking distillation led to more modern cracking processes. It is worthy of note here that it was also recognized that the high-boiling constituents once exposed to cracking were so changed in composition as to be more refractory than the original feedstock. Presumably, incompatibility of these refractory (and highly insoluble) products was also recognized at the time. Whether any effort was made to alleviate the problem is not known.

With the onset of the development of the automobile, the most important part of any refinery became the gasoline-manufacturing facilities. Among the processes that have evolved for gasoline production are thermal cracking, catalytic cracking, thermal reforming, catalytic reforming, polymerization, alkylation, coking, and distillation of fractions directly from crude petroleum.

The problem of how to produce more gasoline from less crude oil was solved by the incorporation of cracking units into refinery operations in which fractions heavier than gasoline were converted into gasoline by thermal decomposition. The early (pre-1940) processes employed for gasoline manufacture were processes in which the major variables involved were feedstock type, time, temperature, and pressure and which need to be considered to achieve the cracking of the feedstock to lighter products with minimal coke formation.

As refining technology evolved, the feedstocks for cracking processes became the residuum or heavy distillate from a distillation unit. In addition, the residual oils and even some of the heavier crude oils were produced as the end-products of distillation processes.

Subjecting these residua directly to thermal processes has become economically advantageous, since on the one hand, the end result is the production of lower-boiling salable materials and on the other hand, the asphaltic materials in the residua are regarded as the unwanted coke-forming constituents. There was also the realization that use of these different feedstocks was yielding products that, although of the same boiling range as from other feedstocks, were chemically estranged from the products from the more conventional crude oils. More refining (usually in the form of hydroprocessing) was required to mitigate the possibility of instability and/or incompatibility.

As the thermal processes evolved and catalysts were employed with more frequency, poisoning of the catalyst with a concurrent reduction in the lifetime of the catalyst became a major issue for refiners. To avoid catalyst poisoning, it became essential that as much of the nitrogen and metals (such as vanadium and nickel) as possible should be removed from the feedstock. The majority of the heteroatoms (nitrogen, oxygen, and sulfur) and the metals are contained in, or associated with, the asphaltic fraction (residuum). It became necessary that this fraction be removed from cracking feedstocks.

With this as the goal, a number of thermal processes, such as tar separation (flash distillation), vacuum flashing, visbreaking, and coking, came into wide usage by refiners and were directed at upgrading feedstocks by removal of the asphaltic fraction.

The method of deasphalting with liquid hydrocarbon is itself a means of initiating incompatibility of the feedstock components by the generation of a "new" feedstock.

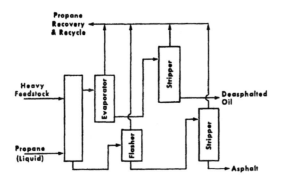

Figure 2.8: Propane deasphalting.

The addition of liquefied gases such as propane, butane, or isobutane (Figure 2.8) induced changes in the liquid medium that made it unable to tolerate the asphaltic constituents, which are precipitated. Thus, deasphalting became a widely used refinery operation in the 1950s and was very effective for the preparation of residua for cracking feedstocks.

61

Table 2.4: Summary of thermal cracking processes.

Visbreaking
    purpose: viscosity reduction
    mild (470-495°C; 880-920°F) heating
    50-200 psi
    low (10%) conversion to <220°C (<430°F)
    heated coil or drum
Delayed coking
    purpose: coke and liquids production
    moderate (480-515°C; 900-960°F) heating
    90 psi
    coke yield: 20-40 wt%
    <430°F yield: 10-30 wt%
    soak drum
Fluid coking
    purpose: liquids production with reduced coke
    make
    severe (480-565°C; 900-1050°F) heating
    10-20 psi
    fluidized bed
    less coke make than delayed coking

Operating conditions in the deasphalting tower depend on the boiling range of the feedstock and the required properties of the product. Generally, extraction temperatures can range from 55 to 120°C (130 to 250°F), with a pressure of 400 to 600 psi. Hydrocarbon:oil ratios of 6:1 to 10:1 by volume are normally used.

3.1 Thermal Cracking

One of the earliest conversion processes used in the petroleum industry was the thermal decomposition of higher-boiling materials into lower-boiling products. This process is known as thermal cracking, although the exact origins of the process are unknown.

The process was developed in the early 1900s to produce gasoline from the "unwanted" higher-boiling products of the distillation process. However, it was soon learned that the thermal cracking process also produced a wide slate of products, varying from highly volatile gases to nonvolatile coke.

The heavier oils produced by cracking are light and heavy gas oil as well as a residual oil that could also be used as heavy fuel oil. Gas oil from catalytic cracking was suitable for domestic and industrial fuel oil or as diesel fuel when blended with straight-run gas oil. The gas oil produced by cracking was also a further important source of gasoline.

The majority of the thermal cracking processes use temperatures of 455-540°C (850-1005°F) and pressures of 100 to 1000 psi. The thermal cracking process using higher-boiling petroleum feedstocks to produce gasoline is now virtually obsolete. There is no guarantee of the chemistry, and the antiknock requirements of modern automobile engines together with the different nature of crude oils (compared with those of 50 or more years ago), has reduced the ability of the thermal cracking process to produce gasoline on an economic basis. Very few new units have been installed since the 1960s, and some refineries may still operate the older thermal cracking units.

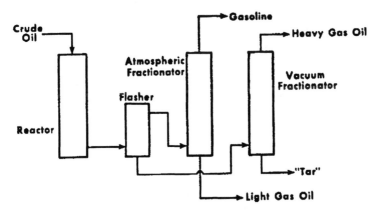

Figure 2.9: The visbreaking process.

3.2 Visbreaking

Visbreaking (viscosity breaking) was initially introduced as a mild thermal cracking operation (Table 2.4) that could be used to reduce the viscosity of residua to allow the products to meet fuel oil specifications.

Alternatively, the visbroken residua could be blended with lighter product oils to produce fuel oils of acceptable viscosity. By reducing the viscosity of the residuum, visbreaking reduces the amount of light heating oil that is required for blending to meet the fuel oil specifications. In addition to the major product, fuel oil, material in the gas oil and gasoline boiling range is produced. The gas oil may be used as additional feed for catalytic cracking units, or as heating oil.

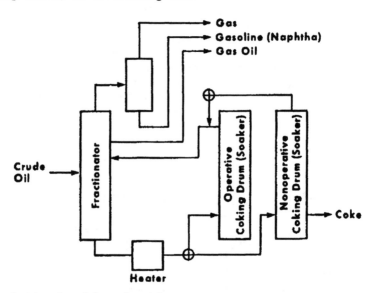

Figure 2.10: The delayed coking process

In a typical visbreaking operation (Figure 2.9), a crude oil residuum is passed through a furnace, where it is heated to a temperature of 480°C (895°F) under an outlet pressure of about 100 psi. The products appear as overhead (distillate) materials and as nonvolatile materials (visbroken bottoms).

Visbreaking is one of those processes that produces a **reactive** product, insofar as the reactions do not terminate at the time of quenching. Instances are known where visbroken bottoms have

64

produced incompatible sediments hours (sometimes days) after the process has been applied.

3.3 Coking

Coking is a thermal process for the continuous conversion of heavy, low-grade oils into lighter products. The feedstock is typically a residuum and the products are gases, naphtha, fuel oil, gas oil, and coke.

After a gap of several years, the recovery of heavy oils caused a renewal of interest in these feedstocks in the 1960s and, henceforth, for coking operations (Speight, 1990).

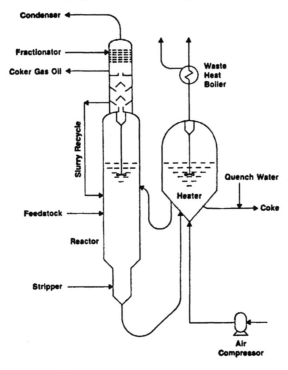

Figure 2.11: The fluid coking process.

Coking processes generally utilize longer reaction times than

PETROLEUM PRODUCTS: INSTABILITY AND INCOMPATIBILITY

Table 2.5: Examples of Evaluation Data for the Estimation of
Product Yields from the Delayed Coking of (a) Wilmington and (b)
East Texas Crude Oil Residua

(a)
Coke, wt.%   = 39.68 - 1.60 x °API
Gas (<C₄), wt.%   = 11.27 - 0.14 x °API
Gasoline, wt. %   = 20.5 - 0.36 x °API
Gas oil, wt. %   = 28.55 + 2.10 x °API

(b)
Coke, wt.%   = 45.76 - 1.78 x °API
Gas (<C₄), wt.%   = 11.92 - 0.16 x °API
Gasoline, wt.%   = 20.5 - 0.36 x °API

the older thermal cracking processes and, in fact, may be
considered to be descendants of those older processes.
Delayed coking is a semicontinuous process (Figure 2.10) in
which the heated charge is transferred to large soaking (or coking)
drums, which provide the long residence time needed to allow the
cracking reactions to proceed to completion. The feed to these
units is normally an atmospheric residuum, although cracked residua
are also used.

The feedstock, including any recycle streams of heavy
products, is heated in a furnace whose outlet temperature varies
from 480 to 515°C (895 to 960°F). The heated feedstock enters one
of a pair of coking drums, where the cracking reactions continue.
The cracked products leave as overheads, and coke deposits form on
the inner surface of the drum. To give continuous operation, two
drums are used; while one is on stream, the other is being cleaned.
The temperature in the coke drum ranges from 415 to 450°C (780 to
840°F) at pressures from 15 to 90 psi.

Overhead products go to the fractionator, where naphtha and
heating oil fractions are recovered. The nonvolatile material is
combined with preheated fresh feed and returned to the reactor.
The coke drum is usually on stream for about 24 hr. before becoming

66

filled with porous coke, after which the coke is removed hydraulically. Normally, 24 hr. are required to complete the cleaning operation and to prepare the coke drum for subsequent use on stream.

There are formulae for estimating product yields from delayed coking (Table 2.5) (Speight, 1981 and references therein), although they vary for individual crude oils and may be of little value now that blends of crude oils are the norm in terms of refinery feedstocks.

However, a major issue arising is that the products from a delayed coker are usually unsaturated and are reactive to oxygen. Sediment and gum formation will occur if the products are not hydrotreated.

Fluid coking is a continuous process that uses the fluidized-solids technique to convert atmospheric and vacuum residua to more valuable products. The residuum is coked by being sprayed into a

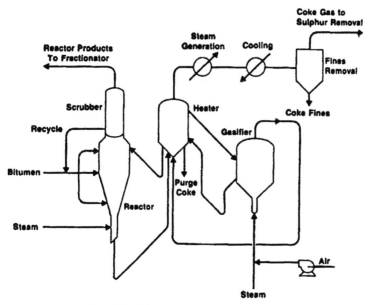

Figure 2.12: The flexicoking process.

Table 2.6: Summary of catalytic cracking processes.

Process parameters
    solid acidic catalyst (e.g.: silica-alumina, zeolite)
    480-540°C (900-1000°F)
    solid-vapor contact
    10-20 psi
    fixed bed/moving bed/fluid bed
Feedstocks
    naphtha
    middle distillate
    gas oil
Products
    C_3-C_4 gases
    low-boiling liquid hydrocarbons
    coke

fluidized bed of hot, fine coke particles, which permits the coking reactions to be conducted at higher temperatures and shorter contact times than can be employed in delayed coking. Moreover, these conditions result in decreased yields of coke; greater quantities of more valuable liquid product are recovered in the fluid coking process.

Fluid coking (Figure 2.11) uses two vessels, a reactor and a burner; coke particles are circulated between these to transfer heat (generated by burning a portion of the coke) to the reactor. The reactor holds a bed of fluidized coke particles, and steam is introduced at the bottom of the reactor to fluidize the bed.

Flexicoking is also a continuous process that is a direct descendent of fluid coking. The unit uses the same configuration as the fluid coker but has a gasification section in which excess coke can be gasified to produce refinery fuel gas.

The flexicoking process (Figure 2.12) is a means by which excess coke-make could be reduced in view of the gradual incursion of the heavier feedstocks in refinery operations. Such feedstocks are notorious for producing high yields of coke (>15% by weight) in thermal and catalytic operations.

3.4 Catalytic Cracking

There are many processes in a refinery that employ a catalyst to improve process efficiency (Speight, 1991). The original incentive arose from the need to increase gasoline supplies in the 1930s and 1940s. Since cracking could virtually double the volume of gasoline from a barrel of crude oil, cracking was justifiable on this basis alone.

In the 1930s, thermal cracking units produced approximately 50% of the total gasoline. The octane number of this gasoline was about 70, compared with 60 for straight-run (distilled) gasoline. The thermal reforming and polymerization processes that were developed during the 1930s could be expected to further increase the octane number of gasoline to some extent, but an additional innovation was needed to increase the octane number of gasoline to enhance the development of more powerful automobile engines.

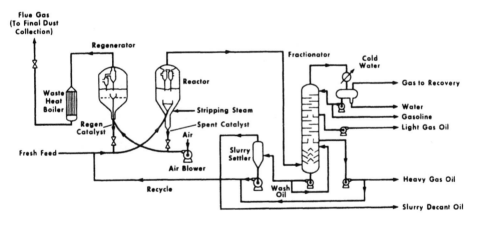

Figure 2.13: The catalytic cracking process.

In 1936 a new cracking process opened the way to higher-octane gasoline; this process was catalytic cracking. This process is basically the same as thermal cracking but differs by the use of a catalyst, which is not (in theory) consumed in the process and directs the course of the cracking reactions to produce more of the desired higher-octane hydrocarbon products.

Catalytic cracking in the usual commercial process (Figure 2.13) involves contacting a gas oil fraction with an active catalyst under suitable conditions of temperature, pressure, and residence time, so that a substantial part (>50%) of the gas oil is converted into gasoline and lower-boiling products, usually in a single-pass operation (Table 2.6).

However, during the cracking reaction, carbonaceous material is deposited on the catalyst, which markedly reduces its activity, and removal of the deposit is vital. This is usually accomplished by burning the catalyst in the presence of air until catalyst activity is reestablished.

The catalyst, which may be an activated natural or synthetic material, is employed in bead, pellet, or microspherical form and can be used as a fixed bed, moving bed, or fluid bed (Speight, 1991; Gray, 1994).

4.0 Hydroprocesses

The use of hydrogen in thermal processes is perhaps the single most significant advance in refining technology. The presence of hydrogen during heating terminates many of the coke-forming reactions and enhances the yields of the lower-boiling components such as gasoline, kerosene, and jet fuel. Moreover, the process produced saturated feedstocks that are resistant to oxidation (and other reactions) and that are usually responsible for the production of sludge, gum, and sediment. However, hydroprocessing may produce a saturated product stream, that is incompatible with another product steam and immiscibility (incompatibility) can result when the two product steams are blended.

Hydrogenation processes for the conversion of petroleum fractions and petroleum products may be classified as **destructive** and **nondestructive**.

Destructive hydrogenation (**hydrogenolysis** or **hydrocracking**) is characterized by the conversion of the higher-molecular-weight constituents in a feedstock to lower-boiling products:

$$R-CH_2-CH_2-R' + H_2 = RCH_3 + R'CH_3$$

Such treatment requires severe processing conditions and the use of high hydrogen pressures to minimize polymerization and condensation reactions that lead to coke formation.

Nondestructive or simple hydrogenation is generally used for the purpose of improving product quality without appreciable alteration of the boiling range:

$$R-N(R')-R'' + 3H_2 = RH + R'H + R''H + NH_3$$
$$R-O-R' + 2H_2 = RH + R'H + H_2O$$
$$R-S-R' + 2H_2 = RH + R'H + H_2S$$
$$R-CH=CH-R' + 2H_2 = R-CH_2-CH_2-R'$$

Mild processing conditions are employed, so that only the more unstable materials are attacked. Nitrogen, sulfur, and oxygen compounds undergo reaction with the hydrogen to remove ammonia, hydrogen sulfide, and water, respectively. Unstable compounds that might lead to the formation of gums or insoluble materials are

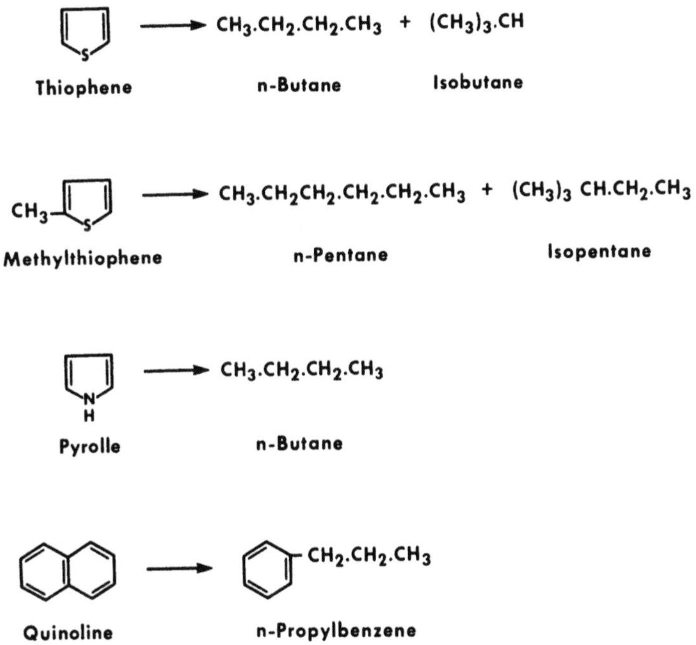

Thiophene $\longrightarrow$ $CH_3.CH_2.CH_2.CH_3$ + $(CH_3)_3.CH$

| Thiophene | n-Butane | Isobutane |

CH_3— $\longrightarrow$ $CH_3.CH_2CH_2.CH_2.CH_2.CH_3$ + $(CH_3)_3 CH.CH_2.CH_3$

Methylthiophene   n-Pentane   Isopentane

$\longrightarrow$ $CH_3.CH_2.CH_2.CH_3$

Pyrolle   n-Butane

$\longrightarrow$ —$CH_2.CH_2.CH_3$

Quinoline   n-Propylbenzene

71

converted to more stable compounds by removal of the heteroelements. There is also the potential that, for the heavier feedstocks, lower temperatures may enhance the yields of liquids and removal of the heteroelements without the formation of the usually high yields of coke (Speight and Moschopedis, 1979).

In actual practice, it is difficult if not impossible to limit the commercial operation to any one particular reaction, but the prevailing conditions may, to a certain extent, minimize one or the other of the reaction options. Nevertheless, the ultimate aim of hydrocracking is to produce as much liquid product as possible. Thus, any hydroprocess that has been designed for application to the heavier oils and residua may require that hydrocracking and hydrotreating occur simultaneously (Speight, 1991).

In a mixture as complex as a residuum or heavy oil, the reaction processes can only be generalized because of difficulties

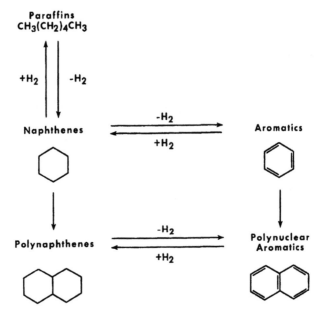

Figure 2.14: Reforming chemistry.

in analyzing not only the products but also the feedstock as well as the intricate and complex nature of the molecules that make up the feedstock. The formation of coke from the high-molecular-weight and polar constituents (i.e., the asphaltenes) of a feedstock (Chapters 13 and 14) is detrimental to process efficiency and to catalyst performance (Skripek et al., 1975; Bunger and Li, 1981; Speight, 1981 and references therein, 1984a, 1984b).

Thus, in the hydrocracking process, attention must also be given to the coke mitigation aspects. For example, in the hydrogen addition options, particular attention must be given to hydrogen management, thereby promoting asphaltene fragmentation to lighter products rather than to the production of coke. The presence of a material with good solvating power, to diminish the possibility of coke formation, is preferred.

In this respect, it is worth noting the reappearance of donor solvent processing of heavy oils (Vernon et al., 1984; McConaghy, 1987), which has its roots in the older visbreaking process hydrogen donor diluent (Carlson et al., 1958; Langer et al., 1961, 1962; Bland and Davidson, 1967).

Hydrotreating is carried out by charging the feed to the reactor, together with hydrogen, in the presence of catalysts such as tungsten-nickel sulfide, cobalt-molybdenum-alumina, nickel oxide-silica-alumina, and platinum-alumina. Most processes employ cobalt-molybdena catalysts which generally contain about 10% of molybdenum oxide and less than 1% of cobalt oxide supported on alumina (Figure 2.14).

The temperatures employed are in the range 300-345°C (570-655°F), while the hydrogen pressures are about 500 to 1000 psi.

Hydrotreating of distillates may be defined simply as the removal of nitrogen-, sulfur-, and oxygen-containing compounds by selective hydrogenation. The hydrotreating catalysts are usually cobalt plus molybdenum or nickel plus molybdenum (in the sulfide) impregnated on an alumina base. The hydrotreated operating conditions are such that appreciable hydrogenation of aromatics will not occur, namely, the pressure is 1000 to 2000 psi, and the temperature is about 370°C (700°F). The desulfurization reactions are usually accompanied by small amounts of hydrogenation and hydrocracking.

Hydrotreating distillates from cracking processes involves a formidable combination of reactions that occur simultaneously:

73

PETROLEUM PRODUCTS: INSTABILITY AND INCOMPATIBILITY

hydrodesulfurization (HDS), hydrodenitrogenation (HDN), and hydrodeoxygenation (HDO). Hydrogenation of aromatics and thermal cracking also occur. In addition, the fate of basic and nonbasic must be distinguished because they have different effects on downstream processes, for example pyrrolic or nonbasic nitrogen has been implicated in sludge formation and light instability of gas oils (Chmielowiec et al., 1987).

To complicate matters, thermal reactions also occur during hydrotreating at ca. 400°C (750°F) and become very active at temperatures in excess of 410°C (770°F), providing the potential for unsaturated species to exist in the product should hydrotreating be inadequate. The consequence, because of the competition for the hydrogen by so many other reactions, is the presence of unsaturated species in the products and, hence, gum formation in a "stable" fuel. The same might be observed when the heteroatom removal reactions fall prey to competing hydrotreating reactions.

Hydrocracking (Table 2.7) is similar to catalytic cracking (Table 2.6), with hydrogenation superimposed and with the reactions taking place either simultaneously or sequentially. Hydrocracking was initially used to upgrade low-value distillate feedstocks, such as cycle oils (highly aromatic products from a catalytic cracker

Table 2.7: Summary of hydrocracking/hydrotreating processes.

Process parameters
    solid acidic catalyst (e.g.: silica-alumina/rare earth metals)
    260-540°C (500-1000°F)
    1000-6000 psi hydrogen
    fixed bed/ebullating bed
Feedstocks
    aromatic naphtha
    recycle oil
    coker oil
Products
    some <C_4 gases
    low-boiling liquid hydrocarbons
    recycle "resid" material

that usually are not recycled to extinction for economic reasons), thermal and coker gas oil, and heavy-cracked and straight-run naphtha.

These feedstocks are difficult to process by either catalytic cracking or reforming, since they usually are characterized by a high polycyclic aromatic content and/or by high concentrations of the two principal catalyst poisons, namely sulfur and nitrogen compounds.

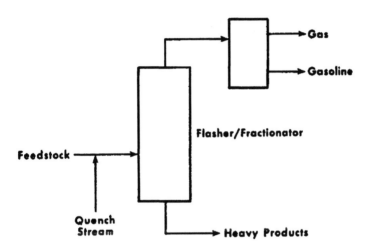

Figure 2.15: Thermal reforming.

5.0 Reforming

When the demand for higher-octane gasoline developed during the early 1930s, attention was directed to ways and means of improving the octane number of fractions within the boiling range of gasoline.

Straight-run (distilled) gasoline frequently had very low octane numbers, and any process that would improve the octane numbers would aid in meeting the demand for higher octane number gasoline. Such a process (called thermal reforming) was developed and used widely but to a much lesser extent than thermal cracking.

Thermal reforming was a natural development from older thermal cracking processes; cracking converts heavier oils into gasoline, whereas reforming converts (reforms) gasoline into higher-octane gasoline. The equipment for thermal reforming is essentially the same as for thermal cracking, but higher temperatures are used.

The chemistry of reforming can be simply expressed as the converse of hydrogenation (Figure 2.15) as cyclization of paraffins as well as aromatization of naphthenes with the concurrent loss of hydrogen. The chemistry may be more complex than this simple concept but the concept does serve to illustrate the changes in character of the hydrocarbon species during reforming. In some cases, it may be these changes in character that lead to instability/incompatibility of the species in a fuel or product.

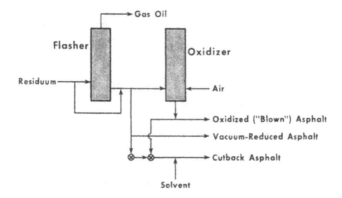

Figure 2.16: Asphalt manufacture.

Thermal reforming is a process (Figure 2.16) in which a feedstock such as 205°C (400°F) end-point naphtha or a straight-run gasoline is heated to 510-595°C (950-1100°F) in a furnace, much the same as a cracking furnace, with pressures from 400 to 1000 psi (27-68 atmospheres). As the heated naphtha leaves the furnace, it is cooled or quenched by the addition of cold naphtha. The material then enters a fractional distillation tower, where any heavy products are separated, and the remainder of the reformed material is separated into gases and reformate. The higher octane of the reformate is due primarily to the cracking of longer-chain paraffins into higher-octane olefins.

Catalytic reforming was commercially nonexistent in the United States prior to 1940 being really a process of the 1950s and showing phenomenal growth in the 1953-1959 time period. Like thermal reforming, catalytic reforming converts low-octane gasoline into high-octane gasoline (reformate). While thermal reforming could produce reformate with research octane numbers of 65 to 80 depending on the yield, catalytic reforming produces reformate with octane numbers of the order of 90 to 95.

Catalytic reforming is conducted in the presence of hydrogen over hydrogenation-dehydrogenation catalysts, which may be supported on alumina or silica-alumina. Depending on the catalyst, a definite sequence of reactions takes place, involving structural changes in the feedstock. This more modern concept actually rendered thermal reforming somewhat obsolete.

Catalytic reformer feeds are saturated (i.e., not olefinic) materials; in the majority of cases that feed may be a naphtha (e.g., coker naphtha). Hydrocracker naphtha that contains substantial quantities of naphthenes is also a suitable feed.

Dehydrogenation is a main chemical reaction in catalytic reforming, and hydrogen gas is consequently produced in large quantities. The hydrogen is recycled though the reactors, where the reforming takes place, to provide the environment necessary for the chemical reactions and also prevents the carbon from being deposited on the catalyst, thus extending its operating life. An excess of hydrogen above whatever is consumed in the process is produced, and as a result, catalytic reforming processes are unique in that they are the only petroleum refinery processes to produce hydrogen as a by-product.

The composition of a reforming catalyst is dictated by the composition of the feedstock and the desired reformate. The catalysts used are principally molybdena-alumina, chromia-alumina, or platinum on a silica-alumina or alumina base. The nonplatinum catalysts are widely used in regenerative process for feeds containing, for example, sulfur, which poisons platinum catalysts, although pretreatment processes (e.g., hydrodesulfurization) may permit platinum catalysts to be employed.

6.0 Isomerization

Catalytic reforming processes provide high-octane constituents in the heavier gasoline fraction, but the normal paraffin components of the lighter gasoline fraction, especially butanes (C_4H_{10}), pentanes (C_5H_{12}), and hexanes (C_6), have poor octane ratings. The conversion of n-paraffins to their branched-chain isomers (isomerization) yields gasoline components of high octane rating in this lower boiling range. Straight-chain paraffins (n-butane, n-pentane, n-hexane) are converted to respective isocompounds (i.e., the respective 2-methyl derivatives but retaining the same total number of carbon atoms) by continuous catalytic (aluminum chloride, noble metals) processes.

Conversion is obtained in the presence of a catalyst (aluminum chloride activated with hydrochloric acid), and it is essential to inhibit side reactions such as cracking and olefin formation.

Present isomerization applications in petroleum refining are used with the objective of providing additional feedstock for alkylation units or high-octane fractions for gasoline blending.

7.0 Alkylation

The combination of olefins with paraffins to form higher isoparaffins is termed alkylation. Since olefins are reactive (unstable) and are responsible for exhaust pollutants, their conversion to high-octane isoparaffins is desirable when possible. In refinery practice, only isobutane is alkylated, by reaction with isobutene or normal butene, and isooctane is the product.

Although alkylation is possible without catalysts, commercial processes use aluminum chloride, sulfuric acid, or hydrogen fluoride as catalysts when the reactions can take place at low temperatures, minimizing undesirable side reactions, such as polymerization of olefins.

Alkylate is composed of a mixture of isoparaffins that have octane numbers that vary with the olefins from which they were made. Butylenes produce the highest octane numbers, propylene the lowest, and pentylenes the intermediate values. All alkylates, however, have high octane numbers (>87), which makes them particularly valuable.

The alkylation reaction as now practiced in petroleum refining is the union, through the agency of a catalyst, of an olefin (ethylene, propylene, butylene, and amylene) with isobutane to yield high-octane branched-chain hydrocarbons in the gasoline boiling range. Olefin feedstock is derived from the gas produced in a catalytic cracker, while isobutane is recovered by refinery gases or produced by catalytic butane isomerization. To accomplish this, either ethylene or propylene is combined with isobutane at 50-280°C (125-450°F) and 300-1000 psi (20-68 atmospheres) in the presence of metal halide catalysts such as aluminum chloride. Conditions are less stringent in catalytic alkylation; olefins (propylenes C_3H_6, butylenes C_4H_8, or pentylenes C_5H_{10}) are combined with isobutane in the presence of an acid catalyst (sulfuric acid or hydrofluoric acid) at low temperatures and pressures (1-40°C; 30-105°F and 14.8-150 psi; 1-10 atmospheres).

8.0 Polymerization

In the petroleum industry, polymerization is the process by which olefin gases are converted to liquid products that may be suitable for gasoline (polymer gasoline) or other liquid fuels. The feedstock usually consists of propylenes (C_3H_6) and butylenes (C_4H_8) from cracking processes or may even be selective olefins for dimer, trimer, or tetramer production.

Polymerization can claim to be the earliest process to employ catalysts on a commercial scale. Catalytic polymerization came into use in the 1930s and was one of the first catalytic processes to be used in the petroleum industry.

Polymerization may be accomplished thermally or in the presence of a catalyst at lower temperatures. Thermal polymerization is regarded as not being as effective as catalytic polymerization but has the advantage that it can be used to "polymerize" saturated materials that cannot be induced to react by catalysts. The process consists of vapor-phase cracking of, for example, propane and butane, followed by prolonged periods at high temperatures (510-595°C; 950-1100°F) for the reactions to proceed to near completion. Olefins can also be conveniently polymerized by means of an acid catalyst. Thus, the treated, olefin-rich feed stream is brought into contact with a catalyst (sulfuric acid, copper pyrophosphate, phosphoric acid) at 150-220°C (300-425°F) and 150-1200 psi (10-81 atmospheres), depending on feedstock and product requirement.

9.0 Other Processes

There are several processing operations worthy of note that

79

are dependent upon the compatibility of asphaltenes and their products.

9.1 Deasphalting

The first such process is the deasphalting process, where asphaltenes, and often resins, are discharged from the feedstock by the addition of hydrocarbon liquids. This is analogous to the laboratory separation procedure, with the exception that the process liquids are often the lower-molecular-weight hydrocarbons liquefied under pressure (Speight, 1991). A similar situation occurs when asphaltenes are deposited on reservoir rock due to the increased solubility of hydrocarbon gases in the petroleum as reservoir pressure increases during maturation (Evans et al., 1971).

9.2 Asphalt Manufacture

Another area where incompatibility might play a detrimental role, during processing or in the product, is in asphalt oxidation, which is an integral part of many asphalt-manufacturing plants (Figure 2.16).

The more polar species in a feedstock will oxidize first in a

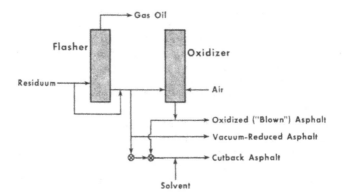

Figure 2.16: Asphalt manufacture.

simulated air-blowing operation (Speight, 1991), and after incorporation of oxygen to a limit, significant changes can occur in asphaltene molecular weight, which is due not so much to oxidative degradation but to the incorporation of oxygen functions that interfere with the natural order of intramolecular structuring.

Should this occur during processing, the result will be a poor-grade asphalt, where phase separation may already have occurred or should it occur in the product, the result can be pavement failure due to a weakening of the asphalt-aggregate interactions.

10.0 References

Bland, W.F., and Davidson, R. L. 1967. Petroleum Processing Handbook, McGraw Hill, New York.

Carlson, C.S., Langer, A.E., Stewart, J., and Hill, R.M. 1958. Ind. Eng. Chem. 50: 1067.

Chmielowiec, J., Fischer, P., and Pyburn, C.M. 1987. Fuel. 66: 1358.

Evans, C.R., Rogers, M.A., and Bailey, N.J.L. 1971. Chem. Geol. 8: 147.

Gray, M.R. 1994. Upgrading Petroleum Residues and Heavy Oils. Marcel Dekker Inc., New York.

Gruse, W.A., and Stevens, D.R. 1960. Chemical Technology of Petroleum. McGraw-Hill, New York.

Hobson, G.D., and Pohl, W. (editors), 1973. Modern Petroleum Technology. Applied Science Publishers Inc., Barking, Essex, England.

Langer, A.W., Stewart, J., Thompson, C.E., White, H.T., and Hill, R.M. 1961. Ind. Eng. Chem. 53: 27.

Langer, A.W., Stewart, J., Thompson, C.E., White, H.T., and Hill, R.M. 1962. Ind. Eng. Chem. Process Design Dev. 1: 309.

McConaghy, J.R. 1987. United States Patent 4,698,147.

McKetta, J.J. (Editor) 1992. Petroleum Processing Handbook. Marcel Dekker Inc., New York.

Nelson, W. L. 1958. Petroleum Refining Engineering. 4th Edition, McGraw-Hill. New York.

Scouten, C.S. 1990. In Fuel Science and Technology Handbook. J.G. Speight (editor). Marcel Dekker Inc., New York.

Speight, J.G. 1981. The Desulfurization of Heavy Oils and Residua. Marcel Dekker Inc., New York.

Speight, J.G. 1990. Fuel Science and Technology Handbook. Marcel Dekker Inc., New York.

Speight, J.G. 1991. The Chemistry and Technology of Petroleum. 2nd Edition. Marcel Dekker Inc., New York.

Speight, J.G. 1994. The Chemistry and Technology of Coal. 2nd Edition. Marcel Dekker Inc., New York.

Speight, J.G., and Francisco, M.A. 1990. Rev. Inst. Fr. Petrole. 45: 733.

Speight, J.G., and Moschopedis, S.E. 1979. Fuel Process Technol. 2: 295.

Speight, J. G. 1984a. in Catalyis in the Energy Scene. S. Kaliaguine and A. Mahay, Eds. Elsevier, Amsterdam. p. 515.

Vernon, L.W., Jacobs, F.E., and Bauman, R.F. 1984. United States Patent 4,425,224.

CHAPTER 3: SOURCES OF LIQUID FUELS. II. COAL

1.0 Introduction

The production of liquid fuels from nonpetroleum sources is not new. For example, the production of liquid fuels from coal has received considerable attention, since the concept does represent alternate pathways to liquid fuels (Figure 3.1) (Cusumano et al., 1978; Anderson and Tillman, 1979; Whitehurst et al., 1980; Speight, 1994 and references cited therein).

Coal has not received the same attention as petroleum as a source of liquid fuels and other products, perhaps because of the

Coal Classification by Rank

Class and Group	Fixed Carbon[a] (%)	Volatile Matter[a] (%)	Heating value[b] (btu/lb)
Anthracitic			
1. Meta-anthracite	>98	<2	-
2. Anthracite	92-98	2-8	-
3. Semianthracite	86-92	8-14	-
Bituminous			
1. Low-volatile bituminous coal	78-86	14-22	-
2. Medium-volatile bituminous coal	69-78	22-31	-
3. High-volatile A bituminous coal	<69	>31	>14,000
4. High-volatile B bituminous coal	-	-	13,000-14,000
5. High-volatile C bituminous coal	-	-	10,500-13,000[c]
Subbituminous			
1. Subbituminous A coal	-	-	10,500-11,500[c]
2. Subbituminous B coal	-	-	9,500-10,500
3. Subbituminous C coal	-	-	8,300-9,500
Lignitic			
1. Lignite A	-	-	6,300-8,300
2. Lignite B	-	-	<6,300

[a]Calculated on dry, mineral-matter-free coal.
[b]Calculated on mineral-matter-free coal containing natural inherent moisture.
[c]Coals with a heating value of 10,500-11,500 btu/lb are classified as high-voltage bituminous coal if they have agglomerating properties and as subbituminous A coal if they are nonagglomerating.
Source: American Society for Testing and Materials, Standard Specifications for Classification of Coals by Rank (ASTM Designation D388-66).

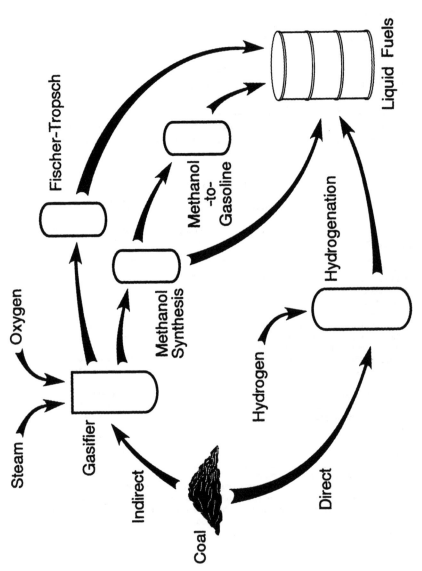

Figure 3.1: General paths for the production of liquids from coal

Table 3.1: Hydrogen Contents of Various Liquid Fuels and a
Bituminous Coal.

Fuel	%C	%H	H/C
Gasoline	86.0	14.0	1.95
Crude oil	85.8	13.0	1.82
Lloydminister heavy oil	83.7	10.9	1.56
Oil sand bitumen (Athabasca)	84.2	10.3	1.47
High-volatile C bituminous coal (Alberta)	77.7	4.9	0.76

overall complexity of the coal system (Gorbaty and Ouchi, 1981; Schobert et al., 1991; Speight, 1994). Its main use, as evidenced by physical property data (Table 3.1) (Speight, 1994), has been in the area of coke production, tars/liquors, and power generation (combustion) as well as the production of gaseous fuels. However, the perception is that at some time in the future there may be the need to use coal to produce a product slate analogous to that obtained from petroleum processing.

The elemental composition of coal exhibited a hydrogen content sufficiently low, especially when compared with other sources of liquid fuels (Table 3.2), that coal was not seriously considered as a source of liquid fuels. The high aromaticity of coal itself (Table 3.3) was perceived as being a barrier to the production of liquid fuels, where low aromaticity (high paraffinicity) was seen as being desirable.

This is not a surprise to those familiar with the older literature, where for decades, coal was used as a source of a considerable number of the lower-molecular-weight hydrocarbon species as chemical intermediates, and what are now referred to as petrochemicals. Indeed, the gasification of coal is an old concept (Nef, 1957; Taylor and Singer, 1957), and conversion of the gases to liquids by way of the Fischer-Tropsch reaction has been commercial for at least 35 years in South Africa (Dry and Erasmus, 1987).

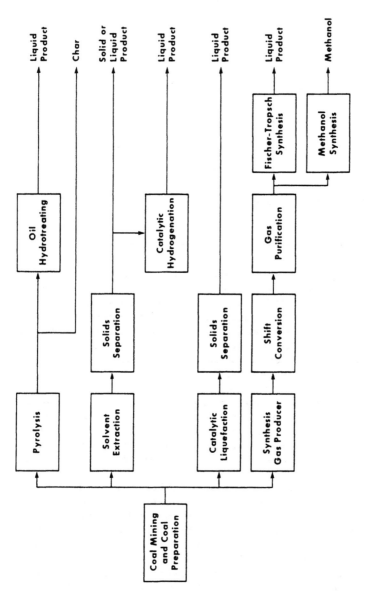

Figure 3.2: Simplified representative of coal liquefacation process types (U.S. ERDA Report 76-67, 1976.)

In fact, the concept of producing liquids from coal (Figure 3.2) is also old (Berthelot, 1869) and is often cited as a viable option for alleviating projected shortages of liquid fuels as well as offering some measure of energy independence for those countries with vast resources of coal who are also net importers of crude oil (Speight, 1994).

Table 3.2   General Summary of Several Pyrolysis and Hydropyrolysis Processes

Process	Reactor type	Reaction temperature		Reaction pressure (psi)
		°C	°F	
Lurgi-Ruhrgas	Mechanical mixer	450-600	840-1110	15
COED	Multiple fluidized bed	290-815	550-1500	20-25
Occidental coal pyrolysis	Entrained flow	580	1075	15
Toscoal	Kiln-type retort vessel	425-540	795-1005	15
Clean coke	Fluidized bed	650-750	1200-1380	100-150
Union Carbide Corp	Fluidized bed	565	1050	1000

Source: Braunstein et al., 1977.

The aim of this chapter is to review the means by which liquids can be produced from coal liquefaction and to outline the general character of these liquids to serve as a reference point for the study of incompatibility phenomena.

Typical Elemental Composition of Peat and Some Representative Coals of Different Rank

Rank	Percent C	Percent H	Percent O	Percent N	Percent S
Peat	55	6	30	1	1.3[a]
Lignite	72.7	4.2	21.3	1.2	0.6
Subbituminous	77.7	5.2	15.0	1.6	0.5
High-volatile B bituminous	80.3	5.5	11.1	1.9	1.2
High-volatile A bituminous	84.5	5.6	7.0	1.6	1.3
Medium-volatile bituminous	88.4	5.0	4.1	1.7	0.8
Low-volatile bituminous	91.4	4.6	2.1	1.2	0.7
Anthracite	93.7	2.4	2.4	0.9	0.6

[a]Ash and moisture content constitute remaining weight percent.

2.0 Liquefaction

There are a number of potential processes for the conversion of coal to liquids (Alpert and Wolk, 1981; Gorin, 1981) (Tables 3.4, 3.5, and 3.6) that have been investigated during the late 1970s and early 1980s (Anderson and Tillman, 1979; Whitehurst et al., 1980; O'Hara, 1981; Speight, 1994), but to bring about full commercialization has not been sufficiently economical. More recent efforts, however, are claiming improved yield of liquid products from the lower-rank coals (Serio et al., 1993), thereby renewing the possibility for the process to be commercially viable.

The first commercial production of liquids from coal was

obtained during carbonization primarily for the production of coke
and gas early in the nineteenth century. Successful research on
coal liquefaction by direct hydrogenation and indirect synthesis
began in the early part of the twentieth century. This culminated
in the production of approximately 100,000 bbl/day (15.9 x 10^6
liters/day) of liquid fuels for the German war effort in the early
to mid-1940s.

While the current objectives are in essence the same, there is
a marked attempt to reduce the operating conditions (ca. 400°C; ca.
750°F; ca. 3000 psi, 20.7 MPa, hydrogen) that were prevalent in the
early work. Furthermore, there are also serious efforts being made
to reduce the hydrogen requirements (an ever-expensive commodity)

Table 3.3   General Summary of Several Solvent Extraction Processes

Process	Reactor	Temperature		Pressure (psi)
		°C	°F	
Consol Synthetic	Stirred-tank	400	750	150-450
Solvent-Refined Coal (SRC)	Plug flow	~ 450	840	1000-1500
Solvent-Refined Lignite (SRL)	Plug flow	370-480	700-895	1000-3000
Costeam	Stirred-tank	375-450	705-840	2000-4000
Exxon Donor Solvent (EDS)	Plug flow	425-480	795-895	1500-2000

Source: Braunstein et al., 1977.

for the process by maximizing the use of the hydrogen in the coal
itself or by the use of a coal-derived solvent that is capable of
donating hydrogen (in situ) to the process (Rudnick and Whitehurst,
1981).

The mild hydrogenation of coal under pressure in the presence
of a solvent carrier or vehicle can also be used to produce a solid
pitch as well as a semisolid asphalt-like material. The degree of
hydrogenation and severity of conditions determine the nature and
range of the product.

The distribution of liquids produced from coal depends on the
character of the coal, on the process conditions, and particularly,
on the degree of "hydrogen addition" to the coal. Hydrocarbon
gases and liquids are produced as by-products and occur
increasingly with severity of reaction. At the limit of severity,

Table 3.4 General Summary of Several Catalytic Liquefaction
Processes

Process	Reactor	Catalyst	Temperature	Pressure (psi)
(a) Catalytic Liquefaction Processes				
H-coal	Ebullated bed	Co-Mo/Al$_2$O$_3$	450(840F)	>3000
Synthoil	Fixed bed	Co-Mo/Al$_2$O$_3$	450(840F)	>2000
CCL	Fixed bed	Co-Mo/Al$_2$O$_3$	400(750F)	2000
Multistage	Expanded bed	Co-Mo/Al$_2$O$_3$	400-430 (750-805F)	1000
(b) Catalytic Hydrogenation Processes				

Bergius	Plug flow	Iron oxide	480(895F)	>3000
University of Utah	Entrained flow	Zinc chloride, tin chloride	500-550 (930-1020F)	>1500
Schroeder	Entrained flow	$(NH_4)_2MoO_4$	500(930F)	2000
"Zinc Chloride"	Liquid phase	Zinc chloride	360-440 (680-825F)	>150

Source: Braunstein et al., 1977.

Table 3.5: Yields of products from Coal Synthetic Crude Oil and Petroleum

	Yield (v/v%)	
Distillation fraction	Synthetic crude[a]	Natural crude[b]
Off-gas	<1	1 - 2
Naphtha:		
Light (IBP-77°C; IBP-170°F)	3	2 - 5
Medium (77-190°C; 170-375°F)	34	12 - 22
Heavy (190-205°C; 375-400°F)	7	1 - 3
Kerosene (205-250°C; 400-480°F)	20	7 - 10
Heavy fuel oil (250-315°C; 480-600°F)	16	11 - 15
Gas oil:		
Light (315-345°C; 600-655°F)	3	5 - 7
Heavy (345-510°C; 655-950°F)	12	18 - 30
Residuum (>510°C; >950°F)	4	10 - 40

[a]Prepared by hydrogenation of Illinois no. 6 coal.
[b]Typical ranges of values expected in natural crude oil.

Table 3.6: Inspections for the feed coal and the product from the Solvent-refined Coal process.

	Feed coal	Product*
Carbon, wt%	71	88
Hydrogen, wt%	5	5
Nitrogen, wt%	1	1.5
Oxygen, wt%	10	3
Sulfur, wt%	3	1
Ash, wt%	7	0.5
Moisture	3	0
Volatile matter, wt%	39	37
Fixed carbon, wt%	51.5	63
Heat content, BTU/lb	12821	15768

*Semi-solid material

Table 3.7: Properties of the products from the Solvent-refined Coal II (SRC II) process.

	Solid fuel	Distillate fuel
Nitrogen, wt%	2.0	0.9
Sulfur, wt%	0.8	0.3
Gravity, °API	-18	5
Boiling range		
°C	425+	
°F	800+	400-800
Viscosity, sus/100°F	-	50
Heat content		
BTU/lb	16000	17300

Table 3.8: Oxygen functions in coal liquids.

Fraction	Function*
Distillate	R-O-R
	Ar-O-Ar
	Ar-OH
Resins	Multifunctional
Asphaltenes	Multifunctional
Weak bases	>NH
Strong bases	>N-
Weak acids	R-OH
	Ar-OH
Strong acids	R-COOH
	Ar-COOH

* R = alkyl or cycloalkyl; Ar = aromatic

Table 3.9: Comparisons of elemental data for asphaltenes from petroleum and from coal liquids.

Property	Petroleum	Coal liquid
Carbon, wt%	82	87
Hydrogen, wt%	8	7
Nitrogen,wt%	1.2	1.4
Oxygen,wt%	1	4
Sulfur, wt%	8	1
Metals, ppm		
vanadium	1200	10
nickel	400	5
H/C atomic ratio	1.2	0.9
Molecular weight		
vpo/benzene	5400	750

methane is the sole hydrocarbon product (hydrogasification). The liquids produced as a result of coal liquefaction processes are different (often referred to as being more complex) than petroleum liquids (Kershaw, 1989; Philp and de las Heras, 1992). In generic terms, the liquids products may be classified as neutral oils (essentially pure hydrocarbons), tar acids (phenols), and tar bases (basic nitrogen compounds).

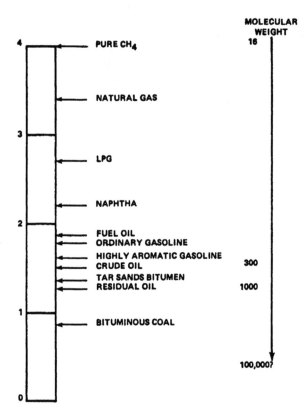

Figure 3.3: Hydrogen content and molecular weights of various fossil fuels.

The neutral oils may make up as much as 80-85% of hydrogenated coal distillate, and as much as 50% of the neutral oils are aromatic compounds, including polycyclic aromatic hydrocarbons. Typical components of the neutral oils are benzene, naphthalene, and phenanthrene (Wood, 1987). Hydroaromatic compounds (cycloparaffins, often called **naphthenes**; see Speight, 1991) are another important component of neutral oils. Hydroaromatic compounds are formed at high hydrogen pressures in the presence of a catalyst, but in the presence of another species capable of accepting hydrogen (such as unreacted coal), hydroaromatic species lose hydrogen to form the thermodynamically more stable aromatic the transfer of hydrogen to unreacted coal during liquid-phase coal compounds and are important intermediates in hydrogenation and solvent refining of coal. The next most abundant component of neutral oil consists of liquid olefins.

The olefins are reactive and cause instability of the liquids by a variety of reactions (ultimately leading to gum formation) that cause changes in the properties of the product with time. On the other hand, olefins are excellent raw materials for the manufacture of synthetic polymers and other chemicals and, thus, can be valuable chemical by-products in coal liquids.

In addition to the neutral oil constituents, coal liquids contain tar acids (consisting of phenolic compounds), which may constitute 5-15% w/w of many coal liquids. They constitute one of the major differences between coal liquids and petroleum.

Petroleum has a much lower content of oxygen-containing compounds, and although tar acids are valuable chemical raw materials, they are not always miscible with hydrocarbon liquid and are troublesome to catalysts in refining processes.

In terms of instability/incompatibility, phenolic compounds have an adverse effect on the storage stability of fuel oil and, by inference, diesel fuel and other fuels (Green et al., 1992). By inference, the addition of coal liquids (containing phenols) to hydrocarbon fuels could have a severely adverse effect on the properties of the fuel The option is some degree of hydrotreating to remove the phenols.

Tar bases containing basic nitrogen make up 2-4% w/w of coal hydrogenation liquids. Tar bases are made up of a variety of compounds, such as pyridine, quinoline, aniline, and higher-molecular-weight analogues.

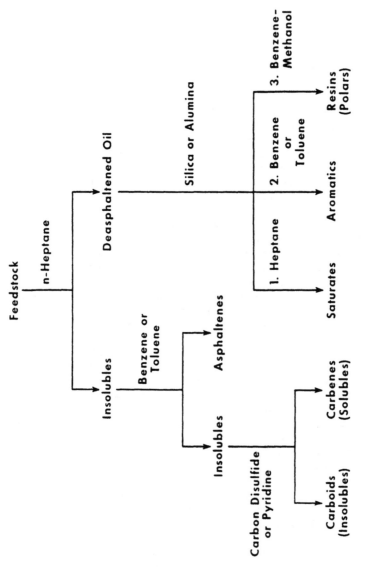

Figure 3.4: Schematic representation of the separation of crude oils and bitumens into asphaltenes, resins, and oil fractions.

96

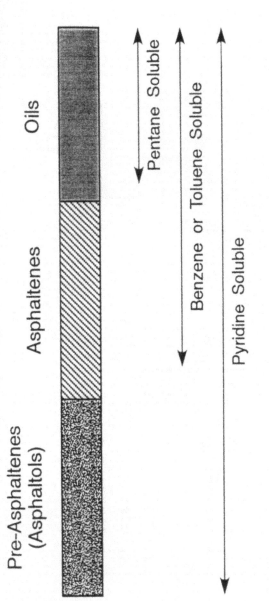

Figure 3.5: Fractions of coal liquids defined by solubility.

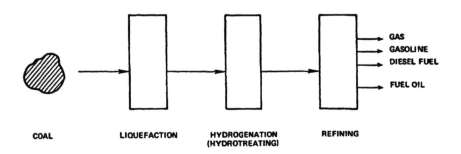

Figure 3.6: Downstream hydrotreating of coal liquids.

Table 3.10: Boiling ranges of various petroleum fractions.

Boiling Ranges of Various Petroleum Fractions

	Boiling range	
Fraction	°C	°F
Fuel gas	-160 - -40	-260 - -40
Propane	-40	-40
Butane(s)	-12 - -1	11 - 30
Light naphtha	-1 - 150	30 - 300
Heavy naphtha	150 - 205	300 - 400
Gasoline	-1 - 180	30 - 355
Kerosene	205 - 260	400 - 500
Stove oil	205 - 290	400 - 550
Light gas oil	260 - 315	500 - 600
Heavy gas oil	315 - 425	600 - 800
Lubricating oil	>400	>750
Vacuum gas oil	425 - 600	800 - 1100
Residuum	>600	>1100

Source: Speight, 1991.

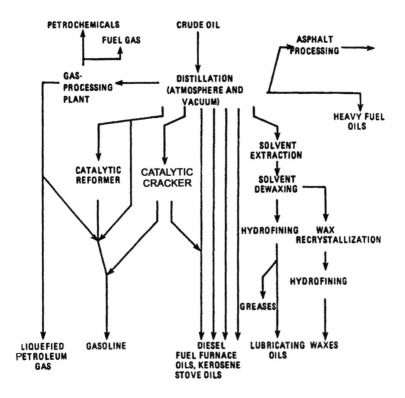

Figure 3.7: Schematic representation of a petroleum refinery.

It is particularly pertinent here to compare hydrogenated coal liquids with crude oil. For example, a coal liquid prepared by hydrogenating Illinois No. 6 coal was distilled, and only the material boiling below 525°C (975°F) was collected. Thus, the coal liquid differed from conventional crude oil in having no heavy ends.

This coal liquid synthetic crude oil had an API gravity of 25.2, specific gravity of 0.9030, pour point of -19°C, Reid vapor pressure of 1.0 psi, viscosity at 15.6°C (60°F) of 37.9 SUS, and sulfur content of 0.13%. These values are not at all out of line compared with typical crude oils. It is interesting that the sulfur content is much lower than that of high-sulfur oil such as Wyoming sour crude (2.9%), or even low-sulfur Louisiana crude (0.38%). However, there are other properties that need to be considered. For example, the aromaticity of the coal liquid needs to be investigated. The functionality of any heteroatom constituents is also worthy of consideration, and there are several other aspects of the properties of the liquid that need to be considered.

The fractional distillation yields of coal-derived synthetic crude oil compared with typical petroleum values (Table 3.7) are useful in determining the utility and refinability of the product. However, it should be noted that the nature of the synthetic crude oil depends on the manner of preparation, and this particular oil should be considered only as an example.

In addition, since this coal-derived crude oil is distilled from a coal hydrogenation mixture, leaving pitch, coke, and other residue behind, it is "unnaturally" deficient in residue compared with petroleum-based crude oil. The distillation yields show that

Table 3.11: Comparisons of elemental data for asphaltenes from petroleum and from coal liquids.

Property	Petroleum	Coal liquid
Carbon, wt%	82	87
Hydrogen, wt%	8	7
Nitrogen,wt%	1.2	1.4
Oxygen,wt%	1	4
Sulfur, wt%	8	1
Metals, ppm		
vanadium	1200	10
nickel	400	5
H/C atomic ratio	1.2	0.9
Molecular weight		
vpo/benzene	5400	750

Table 3.12: Bulk chemicals from coal and their uses.

Product	Chemical constituents	Uses
Aqueous liquor	catechols	adhesives dyes inhibitors (oxidation) pharmaceuticals resins
	phenols	antiseptics disinfectants dyes explosives insecticides
Tar acids	cresols	anti-oxidants insecticides
	xylenols	antiseptics disinfectants dyes explosives insecticides pharmaceuticals weedkillers

Table 3.13: Phenols from coal and their uses.

Compound group*	Uses
phenol	dyes, explosives, pharmaceuticals, perfumes, resins
cresol	dyes, explosives, pharmaceuticals, perfumes, resins, weed killers
xylenol	dyes, oxidation inhibitors, resins,
naphthol	dyes
catechol	oxidation inhibitor, pharmaceuticals, resins, tanning agent
resorcinol	adhesives, antiseptics, dyes, glues, pharmaceuticals, resins

the coal crude is especially rich in the medium naphtha-kerosene fraction boiling up to 250°C (the presence of benzene (b.p. 80°C; 176°F), toluene (b.p. 111°C; 232°F), o-xylene (b.p. 144°C; 291°F), m-xylene (b.p. 139°C; 282°F), p-xylene (b.p. 138°C; 280°F), naphthalene (b.p. 218°C), tetralin (b.p. 207°C; 405°F)), as well as various alkyl (methyl) derivatives of the above compounds, hydroaromatic analogues, and alkanes.

Synthetic crude oil prepared by the hydrogenation of coal is deficient for diesel fuel constituents, since the products in this boiling point range are largely aromatic, which makes the fraction unsuitable for use in a diesel engine.

The olefins in coal synthetic crude oil are reactive and tend to form gum and other undesirable deposits, indicating instability. They may be removed for chemical synthesis or may undergo alkylation processes to form highly branched, high-octane gasoline.

3.0 Carbonization

The thermal decomposition of coal on a commercial scale is often more commonly referred to as carbonization (Holowaty et al., 1981) and is more usually achieved by the use of temperatures up to 1500°C (2730°F). The degradation of the coal is quite severe at these temperatures and produces (in addition to the desired coke) substantial amounts of gaseous products (Howard, 1981).

Carbonization of organic substances is essentially a process for the production of a carbonaceous residue (coke) by thermal decomposition (with simultaneous removal of distillate). The process, which is also referred to as destructive distillation, has been applied to a whole range of organic (carbon-containing) materials, particularly natural products such as wood, sugar, and vegetable matter, to produce charcoal. In the present context, the carbonaceous residue from the thermal decomposition of coal is usually referred to as **coke**, which is physically dissimilar from charcoal and has the more familiar honeycomb-type structure.

However, during coal carbonization, the more familiar coal tar is produced and is isolated by condensation; liquors are also produced. The differentiation between the two is usually in reference to the boiling range. The liquors from carbonization boil in the general range of petroleum naphtha (1-200°C; 30-390°F), often somewhat higher but generally less than 250°C (480°F). The tar from coal carbonization is the higher-boiling material.

The liquor is generally hydrocarbon in nature, highly aromatic, and finds use as a solvent (sometimes referred to as **benzine** in the older or commercial literature). The constituents are of little use for the typical paraffinic liquid fuel but can be used as additives to enhance the octane ratings of fuels. They may resemble, to some extent, the products from reforming processes (Chapter 2), where the reactions leading to aromatization have been carried beyond the usual scope in the refinery. Therefore, the liquors are of some use in a petroleum refinery.

4.0 Indirect Liquefaction

Another category of coal liquefaction processes invokes the concept of the indirect liquefaction of coal; that is, the coal is not converted directly into liquid products.

The indirect liquefaction of coal involves a two-stage conversion operation in which coal is first converted (by reaction with steam and oxygen) to produce a gaseous mixture that is composed primarily of carbon monoxide and hydrogen (syngas; synthesis gas). The gas stream is subsequently purified (to remove sulfur, nitrogen, and any particulate matter), after which it is catalytically converted to a mixture of liquid hydrocarbon products.

In addition to the production of fuel gas, coal may be gasified to prepare synthesis gas (carbon monoxide CO, and hydrogen H_2), which may be used to produce a variety of products, including ammonia, methanol, and liquid hydrocarbon fuels.

The synthesis of hydrocarbons from carbon monoxide and hydrogen (synthesis gas) (the Fischer-Tropsch synthesis) is a procedure for the indirect liquefaction of coal. This process is the only coal liquefaction scheme currently in use on a relatively large commercial scale; South Africa is currently using the Fischer-Tropsch process on a commercial scale in their SASOL complex.

Thus, coal is converted to gaseous products at temperatures in excess of 800°C (1470°F), and at moderate pressures, to produce synthesis gas:

$$C + H_2O = CO + H_2$$

The gasification may be attained by means of any one of

several processes or even by gasification of the coal in place (underground, or in situ, gasification of coal). The exothermic nature of the process and the decrease in the total gas volume in going from reactants to products suggest the most suitable experimental conditions to use in order to maximize product yields.

In practice, the Fischer-Tropsch reaction is carried out at room temperatures of 200-350°C (390-660°F) and at pressures of 75-4000 psi (0.5-4.1 MPa); the hydrogen:carbon monoxide ratio is usually at ca. 2.2:1 or 2.5:1.

Since up to three volumes of hydrogen may be required to achieve the next stage of the liquids production, the synthesis gas must then be converted (by means of the water-gas shift reaction) to the desired level of hydrogen:

$$CO + H_2 = CO_2 + H_2$$

after which the gaseous mix is purified (acid gas removal, etc.) and converted to a wide variety of hydrocarbons:

$$CO + (2n + 1)H_2 = CnH_{2n+2} + nH_2O$$

These reactions result primarily in low- and medium-boiling aliphatic compounds; present commercial objectives are focused on the conditions that result in the production of n-hydrocarbons as well as olefinic and oxygenated materials.

The olefinic materials are hydrogenated to decrease the possibility of instability in the products. The oxygenated materials are usually separated, by design, since they are often alcohols and ethers, which can be used as sources of petrochemicals for manufacture of other products.

5.0 Coprocessing with Other Feedstocks

A more novel aspect of the production of liquids from coal is the use of bitumen and/or heavy oil as process "solvents," and in fact, the coprocessing of coal with a variety of petroleum-based feedstocks (e.g., heavy oils) has received much attention lately (Moschopedis et al., 1980, 1982; Speight and Moschopedis, 1986; Curtis and Hwang, 1992; Rosal et al., 1992; Chakma, 1993; Argirusis and Oelert, 1994).

The concept is a variation of the solvent extraction process for coal liquefaction. Whether the coprocessing option is a means of producing more liquids or whether the coal should act as a scavenger for the metals and nitrogen species in the petroleum material is dependent upon the process conditions. What is certain is that special efforts should be made to ensure the compatibility of feedstocks and products.

Incompatibility can, at any stage of the liquefaction operation, lead to expensive shutdowns as well as to (in respect to the heavier feedstocks) the onset of coke formation (Speight, 1992).

6.0 Refining Coal Liquids and Coal Tars

Coal-derived liquids are those liquids produced from coal by any one of a variety of conversion options (Anderson and Tillman, 1979; Crynes, 1981; Speight, 1994).

At this point, it is necessary to note that in the present context, "conversion" is used to mean the production of liquid products from coal by the application of high temperatures (>350°C) and in some instances, high pressures (>100 psi; >685 kPa). Chemical conversions are excluded from this context.

Before considering the nature of the liquid products from coal, it is worthwhile to consider the various forms in which the heteroatoms (nitrogen, oxygen, and sulfur) are found in coal.

6.1 Heteroatoms

There are several reasons for an interest in the heteroatom types of compounds. First, in terms of the overall structure of coal, it is necessary to know the form in which the heteroatoms may be incorporated into the liquid products.

Second, most of the heteroatoms are initially found in the asphaltene and preasphaltene fractions from the degradation of coal. The materials in these fractions are considered (although there is some disagreement) to be the principal intermediates in the conversion of coal to liquid fuels (liquefaction). The presence of heteroatoms usually has deleterious effects on the storage of coal-derived liquids, since they tend to cause an increase in the viscosity of the liquid products on storage.

Third, the nitrogen- and sulfur-containing compounds are known to poison the catalysts used in the hydrorefining and hydrocracking operations for the upgrading of coal-derived liquids.

105

Fourth, the presence of nitrogen and sulfur in fuels results in high emission levels of the oxides (NO_x, SO_x) of these substances during combustion.

The oxygen content of coal ranges from a high of 20-30% w/w for a lignite to a low of around 1.5-2.5% w/w for an anthracite. The oxygen content (dry basis) of a given coal is normally determined by the difference:

$$\%O = 100 - (\%C + \%H + \%N + \%S + \%ash)$$

Oxygen is known to occur in several different forms in coal, including phenolic hydroxyl, carboxylic acid, carbonyl, ether linkages, and heterocyclic oxygen.

The presence of phenolic hydroxyl, and especially of carboxylic acid, is highly dependent on the rank or degree of coalification. The content of both these groups decreases rapidly with increasing carbon content (rank). Compounds containing the carboxylic acid group ($-CO_2H$) can be very readily extracted from peats and lignites, but seem to disappear in the subbituminous range. Likewise, the methoxy group ($-OCH_3$) appears to be present in the lignitic coals, but has not been shown to be present in significant amounts in the bituminous and higher-rank coals.

The phenolic hydroxyl, ether linkages, and heterocyclic oxygen all appear to be present in bituminous coals and also, to a smaller extent, in the higher-rank coals. The most common form of heterocyclic oxygen is in furan ring systems, and substituted furan rings have been reported in coal extracts, pyrolysis tars, and oxidative degradation products.

However, to exactly what extent the dibenzofuran system exists in the parent coal and to what extent it is formed by the dehydration of phenols during the degradation of the coal matrix is not clear. The isolation of dibenzofuran from oxidative degradation products and low-temperature coal extracts supports the view that it is present in the parent coal.

Nitrogen is typically found in coal in the 0.5-1.5% w/w range. As with oxygen, a number of different types of nitrogen-containing compounds (55) have been isolated from coal-derived liquids. A few examples of these include anilines, pyridines, quinolines, isoquinolines, carbazoles, and indoles. These compounds are frequently substituted with alkyl and aryl groups. The nature of the pyridine and aniline derivatives in the parent coal is a subject of speculation. If the nitrogen heteroatoms were actually an integral part of the original coal matrix, then hydrogenolysis reactions should ultimately result in the cleavage of a number of carbon-carbon bonds, resulting in the formation of a group of methylpyridines and methylanilines.

The sulfur content of coal is quite variable, typically somewhere in the range of 0.5-5.0% w/w. This includes both inorganic (mainly pyrite) and organic sulfur.

A number of classes of organic sulfur compounds have been found in the degradation products of coal, including various derivatives of thiophene, aromatic and aliphatic sulfides, cyclosulfides, thiols, and thiophenols. The alkyl, aryl, and alicyclic thiols are present in coal in relatively small amounts and are normally rather easily removed by hydrogenation of the coal-derived liquid product:

$$R-S-H + H_2 \rightarrow R-H + H_2S$$

The most abundant sulfur functional types found in coal are believed to be derivatives of the thiophene ring system. Other important sulfur types are aryl sulfides, alkyl sulfides, and cyclic sulfides. It should, however, be kept in mind that most of the information on sulfur compounds has been obtained by the analysis of small molecules obtained upon the depolymerization of the coal matrix. The more drastic depolymerization or degradation techniques are very likely to alter the chemical structure of the sulfur compounds originally present in the coal matrix.

6.2 Liquid Products

The liquids derived from coal by high-temperature conversion are complex mixtures of widely differing compound types. From such a broad definition, they may be likened to petroleum, but there the similarities end. Coal-derived liquids are much more acidic than petroleum due to the presence of phenols and have a much narrower molecular weight distribution. However, like petroleum, coal-derived liquids may also be fractionated into a variety of compound types by suitable choice of fractionation techniques. It is at this stage that the asphaltene fraction must be recognized.

Coal-derived asphaltenes, like petroleum asphaltenes (Chapter

107

13), are the fraction of a coal-derived liquid that is insoluble in low-boiling liquid hydrocarbons such as pentane or heptane. Coal-derived asphaltenes are a complex mixture by virtue of their thermal mode of formation from coal.

Current evidence indicates that coal-derived asphaltenes are a collection of predominantly 1 to 4 ring, condensed aromatic systems that contain basic and nonbasic nitrogens as well as oxygen (acidic and etheric) functions. These functionalities play a role in intramolecular relationships within the asphaltene fraction and also with the other constituents of the coal-derived liquid. This latter effect influences the viscosity of the liquid.

Thus, coal-derived asphaltenes are an extremely complex solubility class by virtue of their thermal derivation from coal. It is difficult to accurately depict the structure of coal-derived asphaltenes. A "simple" assessment should include the components of small-ring polynuclear aromatic systems, basic and nonbasic nitrogen, as well as phenolic and etheric oxygen. How these types enjoy an intramolecular existence is another matter, which is difficult and whose definition is left to future research.

The chemistry involved in the liquefaction of coal is complex, but there are indications (Speight, 1994 and references cited therein) that the conversion of coal to liquids involves several steps, including hydrogenation of aromatic rings, decomposition of these rings, cleavage of diphenylmethane-type and 1,2-diphenylethane-type structures, dehydroxylation, and the removal of heteroatoms (nitrogen, oxygen, and sulfur) from ring systems. This is an essentially simple sequence of reactions that, however, lead to a complex distribution of products as reflected in the complex nature of the asphaltene fraction.

It is also worthy of mention here that there is the notion that, whilst coal types are different in general composition, there is a striking similarity in the nature of the major chemical species present in liquids derived from different coal sources. If this is true and if the differences in derived liquid composition can be assigned to differences in the relative amounts of compound types, it may be that the asphaltenes from one coal are similar (except in degree) to the asphaltenes from another coal.

One of the more difficult aspects of identifying the asphaltenic components of coal-derived liquids stems from the spate of processes available for the conversion of coal to liquid products (Speight, 1994). This is, no doubt, similar to the difficulties encountered when attempting to rationalize the structures of petroleum asphaltenes knowing that an asphaltene from one crude may be subtly different from an asphaltene from another crude even though the ultimate analyses may fall into a narrow range (Speight, 1991). Nevertheless one can hope to identify structural features and functionalities that contribute to the phenomenon of "being an asphaltene". Physicochemical and structural studies will aid in resolving these "unknowns" and the potential problems induced in coal-derived liquids by the presence of the asphaltene fraction.

Assessing the structures in coal-derived asphaltenes is difficult because of the variety of thermal options by which they are formed, a fact that must not be ignored. And caution must be taken not to include any thermally labile centers in the hypothetical molecule(s). Nevertheless, it is possible to make reasonable deductions about the structural and functional types present.

The aromatic hydrocarbon systems appear to be more compatible with small-ring aromatics (i.e., are also present) if the molecular fragment containing such systems has had a short residence time in the hot zone; otherwise, thermal aromatization will surely occur. The heteroatom systems have been variously defined on the basis of basic/nonbasic or basic/acidic/neutral groups.

However, the main interests have focused upon the separation of the asphaltenes into basic/nonbasic fractions using the dry hydrogen chloride precipitation technique. Basic pyridine-type nitrogen and acidic (nonbasic) phenol-type oxygen have been identified as the principal heteroatom miscreants in coal-derived asphaltenes because of the detrimental influence of both on downstream processing options.

In general terms, refining coal liquids is often acknowledged to require hydrotreating (deRosset et al., 1979; Speight, 1994 and references cited therein). There has also been the suggestion that fluid catalytic cracking (Chapter 2) also provides a more viable option than hydrocracking for converting coal-derived and coprocessed heavy gas oils to lighter products (Wilson, 1993; Sato et al., 1994).

The very nature (such as the low hydrogen content as well as sulfur and nitrogen contents) (Tables 3.8 and 3.9) of the products

109

from the liquefaction of coal dictates that a refining step be employed as soon as possible after production (Whitehurst et al., 1980; Steedman, 1989; Speight, 1994). The character of the liquid products from coal is extremely complex (Kershaw, 1989). Some of the constituents are miscible with petroleum liquids; others are not, and, the potential for incompatibility.

In addition, the nature of the oxygen functions (Table 3.10) in the raw products also dictates that admixture with petroleum streams before refining can be risky, leading to occurrences of incompatibility. Even after refining, hence, the potential is still real for incompatibility of coal liquids after admixture (blending) with petroleum stocks.

The production of liquid fuels from coal is, in the simplest sense, a means by which hydrogen is added to the coal to improve the atomic hydrogen/carbon ratio and, at the same time, to bring about a reduction in the molecular weight of the product relative to that of the feed, resulting in the production of, at least, the catalytic cracking feedstocks and, ultimately, liquid fuels (Figure 3.3).

Liquid products from coal are generally different from those produced by petroleum refining, particularly as they can contain substantial amounts of phenols. Therefore, there will always be some question about the place of coal liquids in refining operations. For this reason, there have been some investigations on the "next-step" processing of coal liquids (Steedman, 1989) and liquids from coal/bitumen coprocessing (Wilson, 1993).

6.3 Asphalt Products

The mild hydrogenation of coal under appropriate conditions to produce bitumen-like materials requires some comparison and evaluation with petroleum asphalt materials.

Petroleum-derived asphalt is the product of processing the nonvolatile constituents of petroleum and can be generally defined in terms of asphaltenes, resins, and oil constituents (Figure 3.4) (Speight, 1991). The asphaltenes (Chapter 13) are considered to impart the plastic properties generally associated with asphalt. These constituents represent various stages in an oxidation or blowing process; for instance, air blowing at elevated temperatures will convert resins to asphaltenes, and further reaction leads to decomposition and the formation of coke.

110

The chemical nature of asphalt is generally different from that of the synthetic bitumen-like materials derived from coal hydrogenation. But at the same time there may be enough similarity to consider that asphaltene-like materials may be dispersed in oil constituents of varying aromaticity, subject to the conditions of performance, and may be further treated to meet specifications.

However, a difficulty encountered in evaluating and comparing asphalt properties lies in devising and correlating elemental laboratory tests to actual road performance for these non-Newtonian substances. The range of usual tests, such as shear viscosity and viscosity-temperature-susceptibility, has not been entirely adequate.

In general, asphaltenes provide the major element of viscosity, and viscosity-temperature susceptibility increases with viscosity. These effects are contradictory, as some statements are found to the effect that asphaltenes decrease temperature susceptibility and give high fluidity flow viscosity. In fact, it seems to be a predominant opinion that asphaltenes lower the temperature susceptibility and that by increasing the asphaltene content (at least up to a point), the temperature susceptibility will be improved. This depends on many "ifs" and such qualifying phrases as "other things being equal."

While it is conceded that a bitumen-like product obtained from the hydrogenation of coal is not the same as a petroleum asphalt, still there are similarities that may permit characterization in similar terms.

The same element of immiscibility exists between oils and asphaltenes or tars (oxygenated materials). The bitumen-like products and coal liquids contain volatile oils and asphaltene-like constituents (Figure 3.5). However, the asphaltenes are unlike their petroleum counterparts in terms of aromaticity and heteroatom functionality (Table 3.11).

Thus, incompatibility with hydrocarbon liquids might be anticipated and is often realized. However, in terms of the coal liquids themselves, the oils are predominantly aromatic, giving stable sol-like dispersions with high-temperature susceptibilities and exhibiting Newtonian behavior.

Such a bitumen lends itself to a variety of modifications and uses. Steam distillation under vacuum, a universal practice in asphalt processing, would remove volatile materials, and the nonvolatile residue could be air blown to an asphalt. The residue from distillation, if pitch-like, could be fluxed with a nonvolatile oil and excessive asphaltene constituents could be precipitated by solvent refining.

6.4 Refining

The means by which hydrogenation can occur will vary from process to process and may even occur as part of the process by the use of a hydrogen atmosphere and a solvent capable of donating hydrogen to the system and the type of catalyst employed. Nevertheless, in the more general sense, at some stage of the operation, the liquid products need to be stabilized (i.e., freed from unsaturated materials as well as nitrogen, oxygen, and sulfur species) by what may be simply referred to as a hydrotreating operation.

In the simplest sense, this operation may be viewed as occurring downstream of the liquefaction process (Figure 3.6).

Current concepts for refining the products of coal liquefaction processes rely for the most part on the already existing petroleum refineries, although it must be recognized that the acidity (i.e., phenol content) of the coal liquids and the potential incompatibility of the coal liquids with conventional petroleum (or even heavy oil) feedstocks may pose severe problems within the refinery system.

The first essential step in refining coal liquids is severe catalytic hydrogenation to remove most of the nitrogen, sulfur, and oxygen and to convert at least part of the high-boiling material to lower-boiling distillates that might be further refined. This is analogous to the hydrodesulfurization of heavy oils using a preliminary cracking technique, so that after product separation (by distillation), the most suitable choice of process conditions can be made (Speight, 1981).

However, a major limiting factor in refining coal liquids is due to the high aromatics content and to the condensed nature of many of the aromatic ring systems. Thus, to produce liquid fuels of the types currently in demand, each condensed aromatic ring would have to be hydrogenated (saturated) and cracked to produce the lower-boiling distillate material.

The hydrogen demand for such conversions and the effect of these polynuclear aromatic systems (especially those that contain nitrogen and other heteroatoms) on catalysts are very worthy hurdles to overcome! Nevertheless, they are hurdles that can be surpassed, and by a variety of process conditions.

A petroleum refinery involves various operations (Chapter 2), and it is the intent of this section to present a brief outline of a petroleum refinery to indicate which methods may be available to upgrade to liquefaction products.

While appearing to be a relatively facile system, the petroleum refinery is actually a complex integrated series of operations (Figure 3.7) that ultimately results in the production of high-value, salable materials from low-value feedstocks (Speight, 1991).

The operation can vary from the relatively "simple" distillation (or "skimming") process to the much more complex thermal conversion units in which the crude feedstocks are thermally degraded to lighter (lower-molecular-weight) marketable products.

Processes involving the use of a variety of complex and expensive catalysts are also a necessary part of any refinery, and such processes will play an important role in the processing of the products from coal liquefaction units.

In the catalytic cracking process, the object is to produce gasoline, heating oil, and the like from a heavier feedstock (such as a gas oil) (Table 3.12) by means of an aluminosilicate base catalyst. However, the reactions that occur are varied, and especially with the heavier or more aromatic feedstocks, there is the inevitable deposition of carbon (coke) on the catalyst and the accompanying decrease in catalyst activity.

At this point a comparison of asphaltenes (the predominant coke formers in a feedstock) from petroleum with those isolated from coal liquids is worthwhile (Table 3.13). The comparisons should be borne in mind when processability comparisons are necessary.

Hydrocracking is a process that accomplishes the same goals as catalytic cracking, but the presence of hydrogen and more specific catalysts often allows a much better control of the reaction and, therefore, results in a better distribution of products. The hydrocracker is operated at elevated pressures (several thousand psi in the case of the heavier feedstocks; 1 psi = 6.895 kPa) and employs a bifunctional catalyst that has sites capable of promoting the hydrogenation reactions as well as the cracking reactions. Thus, while current refinery technology may suffice to a point for the refining of the liquid and solid products from liquefaction processes, there are many aspects of the operation that may need some modification when the products from coal become a major refinery feedstock. Of course, this may dictate the creation and evolution of a completely new refining technology.

6.5 Coal Tar

The carbonization of coal to produce "coal gas" for street and house lighting in the closing years of the eighteenth century also produced quantities of tar, which (during the following 50 years) were mostly discarded as a troublesome and unnecessary by product.

However, the development of a western European chemical industry brought increasing importance to coal tar as a source of the precursors that were to be used for the synthesis of dyes as well as raw materials for the production of solvents, pharmaceutical products, synthetic fibers, and plastics (Tables 3.14 and 3.15). Coal tar can also be upgraded to gasoline and other liquid fuels (Juntgen et al., 1981; Speight, 1994).

In fact, in the manner of crude petroleum, most high-temperature coal tars are fractionated by distillation into light oil, middle (or tar acid) oil, and heavy (or anthracene) oil. This primary separation is carried out by means of batch stills (vertical or horizontal; capacity of 3000-8000 U.S. gallons; 11-30 x 10^3 liters), or by means of continuous "pipe" stills in which the tar is heated to a predetermined temperature before injection into a fractionating tower.

The light oil fraction, consisting of naphtha (72%), toluene (11-19%), xylene (3-8%), styrene (1-1.5%), and indene (1-1.5%), is processed either into gasoline and aviation fuel components or is fractionated further to provide solvents and petrochemical feedstocks.

In either case, upgrading involves removal of sulfur compounds, nitrogen compounds, and unsaturated materials. This is usually accomplished by acid washing in batch agitators or by hydrogenation over a suitable catalyst (e.g., cobalt-molybdenum or nickel-tungsten on a support).

In the acid wash, the crude material is mixed with strong sulfuric acid, neutralized (with ammonia liquor or caustic soda), and after separation of the aqueous phase, steam distilled or stripped of higher-molecular-weight material by centrifuging. The hydrogenation process conditions vary with the nature of the material to be removed (e.g., sulfur or olefinic material) but could typically be 300-400°C (570-750°F) and 500-1500 psi (3.4-10.3 MPa) hydrogen.

The middle oil usually boils over the range 220-375°C (430-710°F), and after extraction of the tar acids, tar bases, and naphthalene, can be processed to diesel fuel, kerosene, or creosote. In this context, it should be noted that the tar acids, which are mostly phenol, cresols, and xylenols, can be recovered by mixing the crude middle oils with a dilute solution of caustic soda, separating the aqueous layer, and passing steam through it to remove residual hydrocarbons. The acids are then recovered by treatment of the aqueous extract with carbon dioxide or with dilute sulfuric acid and are then fractionated by distillation in vacuo.

Tar bases are isolated by treating the acid-free oil with dilute sulfuric acid, and the bases are regenerated from the acid solution by addition of an excess of alkali (e.g., caustic soda or lime slurry). The mixture is then fractionated to produce pyridine, picoline, lutidine, aniline, quinoline, and isoquinoline.

The temperature to which the distillation of the heavy oil fraction is taken depends on the type of residue (pitch) that is desired but usually lies within the range 450-550°C (840-1020°F). In all cases, the distillate is an excellent source of hydrocarbons, such as anthracene, phenanthrene, acenaphthene, fluorene, and chrysene.

The residual coal tar pitches are complex mixtures that contain several thousand compounds (mostly condensed aromatic compounds) and may, by analogy, be likened to the vacuum residua that are produced in a petroleum refinery and represent the materials in petroleum that have boiling points in excess of 565°C (1050°F).

7.0 References

Alpert, S.B., and Wolk, R.H. 1981. In Chemistry of Coal Utilization. Second Supplementary Volume. M.A. Elliott (editor). John Wiley & Sons Inc., New York. Chapter 28.

American Society for Testing and Materials. 1993. Standard Specifications for Classification of Coals by Rank. ASTM Designation D386-66. Philadelphia, Pennsylvania.

Anderson, L.L., and Tillman, D.A. 1979. Synthetic Fuels from Coal: Overview and Assessment. John Wiley & Sons Inc., New York.

Argirusis, C., and Oelert, H.H. 1994. Erdoel Kohle Erdgas Petrochem. 46: 454.

Berthelot, M. 1869. Bull. Soc. Chim. Fr. 11: 278.

Braunstein, H.M., Copenhaver, E.D., and Pfuderer, H.A. 1977. Environmental, Health, Control Aspects of Coal Conversion. Volume 1. Oak Ridge National Laboratory, Oak Ridge, Tennessee.

Chakma, A. 1993. Fuel Process. Technol. 36: 147.

Crynes, B.L. 1981. In Chemistry of Coal Utilization. Second Supplementary Volume. M.A. Elliott (editor). John Wiley & Sons Inc., New York. Chapter 29.

Curtis, C.W., and Hwang, J.-S. 1992. Fuel Process. Technol. 30: 47.

Cusumano, J.A., Dalla Betta, R.A., and Levy, R.B. 1978. Catalysis in Coal Conversion. Academic Press Inc., New York.

deRosset, A.J., Tan, G., and Gatsis, J.G. 1979. In Refining of Synthetic Crudes. M.L. Gorbaty and B.M. Harney (editors). Advances in Chemistry Series No. 179. American Chemical Society, Washington, DC. Chapter 7.

Dry, M.E., and Erasmus, H.B. de W. 1987. Annu. Rev. Energy. 12: 1.

Gorbaty, M.L., and Ouchi, K. (Editors). 1981. Coal Structure. Advances in Chemistry Series No. 192. American Chemical Society, Washington, DC.

Gorin, E. 1981. In Chemistry of Coal Utilization. Second Supplementary Volume. M.A. Elliott (editor). John Wiley & Sons Inc., New York. Chapter 27.

Green, J.B., Stirling, K.Q., and Ripley, D.L. 1992. Proceedings. 4th International Conference on the Stability and Handling of Liquid Fuels. Volume 1. Reports DOE/CONF-911102. Washington, DC. P. 503.

Holowaty, M.O., Barrett, R.E., Giammar, R.D., and Hazard, H.R.

1981. In Coal Handbook. R.A. Meyers (editor). Marcel Dekker Inc., New York. Chapter 9.

Howard, J.B. 1981. In Chemistry of Coal Utilization. Second Supplementary Volume. M.A. Elliott (editor). John Wiley & Sons Inc., New York. Chapter 12.

Juntgen, H., Klein, J., Knoblauch, K., Schroter, H.J., and Schulze, J. 1981. In Chemistry of Coal Utilization. Second Supplementary Volume. M.A. Elliott (editor). John Wiley & Sons Inc., New York. Chapter 30.

Kershaw, J. (editor). 1989. Spectroscopic Analysis of Coal Liquids. Elsevier, Amsterdam.

Moschopedis, S.E., Hawkins, R.W., Fryer, J.F., and Speight, J.G. 1980. Fuel. 59: 647.

Moschopedis, S.E., Hawkins, R.W., and Speight, J.G. 1982. Fuel Process. Technol. 5: 213.

Nef, J.U. 1957. In A History of Technology. Volume 3. E.J. Holmyard, A.R. Hall, and T.I. Williams (editors). Clarendon Press, Oxford, England. Chapter 3.

O'Hara, J.B. 1981. In Coal Handbook. R.A. Meyers (editor). Marcel Dekker Inc., New York. Chapter 11.

Philip, R.P., and de las Heras, F.X. 1992. In Chromatography. Part B: Applications. 5th Edition. E. Heftmann (editor). Elsevier, Amsterdam. Chapter 21.

Rosal, R., Caboo, L.F., Dietz, F.V., and Sastre, H. 1992. Fuel Process. Technol. 31: 209.

Rudnick, L.R., and Whitehurst, D.D. 1981. In New Approaches in Coal Chemistry. B.D. Blaustein, B.C. Bockrath, and S. Friedman (editors). American Chemical Society, Washington, DC. Chapter 9.

Sato, Y., Yamamoto, Y., Kamo, T., and Miki, K. 1994. Sekiyu Gassaishi. 37: 58.

Schobert, H.H., Bartle, K.D., and Lynch, L.J. (editors). 1991. Coal Science II. Symposium Series No. 461. American Chemical Society, Washington, DC.

Serio, M.A., Kroo, E., Charpenay, S., and Solomon, P.R. 1993. Div. Fuel Chem. Am. Chem. Soc. 38(3): 1021.

Speight, J.G. 1981. The Desulfurization of Heavy Oils and Residua. Marcel Dekker Inc., New York. Chapter 6.

Speight, J.G. 1991. The Chemistry and Technology of Petroleum. Second Edition. Marcel Dekker Inc., New York.

Speight, J.G. 1992. Proceedings. 4th International Conference on the Stability and Handling of Liquid Fuels. Volume 1. Report DOE/CONF-911102. U.S. Department of Energy, Washington, DC. P. 169.
Speight, J.G. 1994. The Chemistry and Technology of Coal. 2nd Edition. Marcel Dekker Inc., New York.
Speight, J.G., and Moschopedis, S.E. 1986. Fuel Process. Technol. 13: 215.
Steedman, W. 1989. In Spectroscopic Analysis of Coal Liquids. J. Kershaw (editor). Elsevier, Amsterdam. Chapter 3.
Taylor, F.S., and Singer, C. 1957. In A History of Technology. Volume 2. E.J. Holmyard, A.R. Hall, and T.I. Williams (editors). Clarendon Press, Oxford, England. Chapter 10.
Whitehurst, D.D., Mitchell, T.O., and Farcasiu, M. 1980. Coal Liquefaction: The Chemistry and Technology of Thermal Processes. Academic Press Inc., New York.
Wilson, M.F. 1993. Div. Fuel Chem. Am. Chem. Soc. 38(3): 1058.

CHAPTER 4: SOURCES OF LIQUID FUELS. III. OIL SHALE

1.0 Introduction

Oil shale is, in the simplest sense, a rock, and it does not contain any oil! In more detailed terms, oil shale is a fine-textured sedimentary rock that contains organic matter from which oil can be liberated by heating (Russell, 1980; Allred, 1982; Yen, 1976; Scouten, 1990).

The organic matter is referred to as **kerogen** which has a varying composition (Scouten, 1990) and is insoluble in organic solvents (Durand, 1980).

There is, on occasion, soluble material that can be extracted from oil shale using organic solvents. This material is referred to as **bitumen,** but it should not be confused with the bitumen that is defined as a naturally occurring heavy oil (Chapter 1).

Different types of kerogen can be recognized by examination and can also be characterized by their respective maturation paths on the van Krevelen (H/C versus O/C) diagram (Scouten, 1990). Conversion of kerogen to oil (**shale oil**) can vary from a low of 10% w/w to yields as high as 70% w/w.

There does appear to be a relationship between the aliphatic carbon in the kerogen and the yield of shale oil (Figure 4.1), but it cannot be presumed that all of the aliphatic material is converted to oil and that all of the aromatic material is converted to carbon (deposited on the mineral matrix). However, in very general terms, oil yield can be considered to be a function of the aliphatic nature of the kerogen (Figure 4.2).

The attractiveness of oil shale to the oil industry is, in the simplest sense, due to the composition (Sinor, 1988). The organic constituents of oil shale, on the other hand, have hydrogen to carbon atomic ratios similar to those found in petroleum and have only small proportions of organic oxygen. Furthermore, liquid products (shale oil) obtained by thermal means from oil shale typically have an atomic hydrogen:carbon ratio of ca. 1.35.

In spite of the potential for added-value products, most oil shale companies have tended to concentrate on producing a single product stream. In short, there has been the tendency to compete with petroleum by producing a crude oil substitute. Unfortunately, whenever a shale oil industry was established strictly to compete

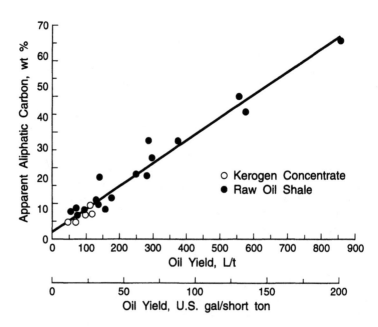

G.E. Maciel, V.J. Bartuska and F.P. Miknis, Fuel, **58**, 155 (1979);
Ibid. **57**, 505 (1978)

Figure 4.1: Carbon-type content and oil yield for various oil shales.

head-to-head with petroleum, the industry became susceptible to the periodic downward fluctuations in petroleum prices.

In only a few instances outside of Scotland, the industry was large enough or possessed enough technical capability to undertake the development of alternative product slates. The Russian industry has long viewed liquid shale oil products not only as an alternative fuel but also as a chemical feedstock. This is at least partly due to two facts: (1) shale retorting was established for the major purpose of producing gas, and the tars were only a

OIL YIELD IS A FUNCTION OF H/C RATIO

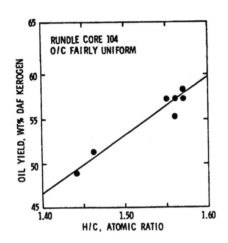

Figure 4.2: Aliphatic character of various kerogen samples.

co-product or by-product; (2) the richness of the resource allowed direct combustion in boilers and power plants where otherwise an oil fuel might have been needed. Thus, Russia has considered petrochemical and other nonfuel product applications shale oil to a far greater extent than any other country (Figure 4.3).

The term of oil shale, like the terms tar sand and oil sand, is

Table 4.1: Approximate composition of kerogen.

C:	69-82
H:	7-11
N:	1- 3
O:	5-17
S:	1- 8

a misnomer (Chapter 1). Just as the material deposited on the sand is not tar and is not oil (rather a heavy bituminous material; Chapter 1), the material in oil shale is not oil. It is more generally called **kerogen**, which is a derivation of an expression meaning "oil generating."

However, like coal and unlike petroleum, kerogen has a range of elemental compositions (Table 4.1), which tends to be a disadvantage in the generation of shale oil, since the products can also vary in elemental composition and fractional composition, analogous to coal liquids. Such variation in composition not only adds to the complexity of the refining operations but also increases the potential for incompatibility to occur when the various liquid streams are blended.

At this time, economic and commercial considerations do not

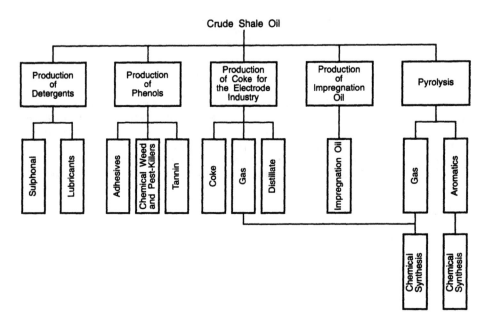

Figure 4.3: Processing scheme for shale oil.

encourage the use of oil shale as an alternative source for the production of liquid fuels, at least not for the near future. However, the use of oil shale derivatives for other products, such as asphalt modifiers, promises to give new life to the on-again,off-again oil shale industry (Scouten, 1990).

The hesitancy of industry to commercialize oil shale technology may be overcome, since a measure of energy independence is truly desired by those countries that are deficient in petroleum resources but rich in oil shale resources. A similar argument is used for development of coal resources as sources of liquid fuels. Coal does have other uses (Chapter 3), which have led to increased mining activities in countries rich in coal resources.

Thus, it is not surprising that there are many investigations of the quality of fuels that could be derived from such alternative energy sources, the effect of the various production methods and processing conditions on the quality of the fuels, and the means for improving their properties. Property improvement is a key issue in the development of liquid fuels (and other products of interest), since the stabilities of these products are generally inadequate.

There are several reasons for the interest in the quality of fuel products derived from alternative sources such as oil shale. First is the endeavor to prepare the necessary technologies in case they would be needed. Second is the fact that the investigation of this complex subject is of great interest and helps in the understanding of the processes associated with fuels of unconventional compositions.

Since shale oils and fuels derived from oil shales are of an extremely poor quality in terms of stability and utilization, emphasis will be given in this chapter to the character of the products and the treatment necessary to produce usable materials. At this time, the potential for a separate refinery to process only oil shale liquids can be likened to the potential for a refinery to process only coal liquids. It is considered more likely that refining of oil shale products would occur in a petroleum refinery, where hydroprocesses are available to remove olefinic materials (Hisamitsu et al., 1993). There would, however, need to be modifications of part of the operations to accommodate the oil shale products. Thus, there is also the observation of the potential incompatibility with other liquid fuels and products from

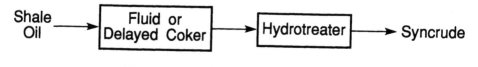

Advantages

- Built In Departiculation
- 1050⁺ Destruction
- Molecular Weight Reduction
- Easier Hydrotreating

Disadvantages

- Yield Losses to Gas and Coke

Figure 4.4: Processing options for shale oil - delayed coking and hydrotreating.

petroleum.

Hence, the focus of this chapter is the generation of products from oil shale, treatment of the products, and their potential as liquid fuels.

2.0 Liquid Fuels from Oil Shale

One of the main items at this point is to provide an understanding about the type and extent of the required treatment and refining processes to which the shale oil and/or the oil shale derived products have to be subjected, prior to their being blended with conventional fuel products (Table 4.2).

The premise is that shale oils and oil shale derived products have to be not only treated, but that the treated products, not being of a sufficiently satisfactory quality, have also to be blended with the corresponding crude oil fractions or crude oil products.

Crude shale oil, sometimes termed retort oil, is the liquid oil condensed from the effluent in oil shale retorting. Crude shale oil typically contains appreciable amounts of water and

Table 4.2: Challenges for oil shale processing.

• Particulates	-	Plugging on processing
	-	Product quality
• Arsenic content	-	Toxic
	-	Catalyst poison
• High pour point	-	Oil not pipeline quality
• Nitrogen content	-	Catalyst poison
	-	Contributes to instability
	-	Toxic
• Diolefin	-	Contributes to instability
	-	Plugging on processing

solids, as well as having an irrepressible tendency to form sediments. As a result, it must be upgraded to a synthetic crude oil (syncrude) before being suitable for pipelining or substitution for petroleum crude as a refinery feedstock. However, shale oils are sufficiently different from petroleum crudes that processing shale oil presents some unusual problems.

Retorting is the process of heating oil shale in order to recover the organic material, predominantly as a liquid. To achieve economically attractive recovery of a product, temperatures of 400-600°C (750-1100°F) are required. Therefore, to avoid wasteful combustion of the products, a retort is simply a vessel in which the oil shale is heated and from which the product gases and vapors can escape to a collector.

Shale oils, especially those from certain oil shale where the kerogen might have a high nitrogen content, have particularly high nitrogen contents, usually of the order of 1.7-2.2 wt% compared with a nitrogen content of 0.2-0.3 wt% for a typical petroleum.

In many other shale oils, nitrogen contents might be lower but are still higher than those typical of petroleum. Because retorted

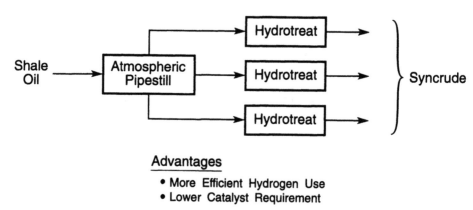

Figure 4.5: Processing options for shale oil - distillation and hydrotreating.

shale oils are produced by a thermal cracking process, olefin and diolefin contents are high.

It is the presence of these olefins and diolefins, in conjunction with high nitrogen contents, that gives crude shale oils their characteristic instability toward sediment formation. The sulfur contents of shale oils vary widely but are generally lower than those of high-sulfur petroleum crudes and tar sand bitumen.

In addition to the olefins and diolefins mentioned above, some shale oils contain appreciable amounts of aromatics, polar aromatics, and pentane insolubles (asphaltenes). Oxygen contents are higher than those typically found in petroleum, but lower than those of crude coal liquids. Crude shale oils also contain

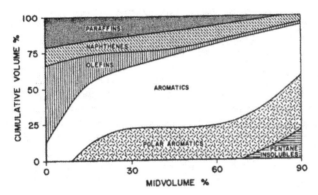

Figure 4.6: Hydrocarbon distribution in shale oil.

appreciable amounts of soluble arsenic, iron, and nickel that cannot be removed by filtration.

In spite of the variation and quantity of the heteroatom species in oil shale, what is perhaps exceptional about shale oil is that it contains a large variety of hydrocarbon compounds (Table 4.3) including paraffins, cycloparaffins, olefins, aromatics, furans, thiophenes, hydroxy-aromatics, pyrroles, and pyridines. Minor constitutents such as thiophenofurans, hydroxy-

Table 4.3: Major compound types in shale oil.

Saturates	Heteroatom systems
paraffins	benzothiophenes
cycloparaffins	dibenzothiophenes
Olefins	phenols
Aromatics	carbazoles
benzenes	pyridines
indans	quinolines
tetralins	nitriles
naphthalenes	ketones
biphenyls	pyrroles
phenanthrenes	
chysenes	

127

nitrogen compounds, ketones, aldehydes, acids, amides, nitriles, and dinitrogen compounds are also present. Alkyl substituents of the compounds can range from one to 50 carbon atom chains. Many of these constituents are potentially incompatible with petroleum.

However, the concentrations of individual constituents can vary substantially and are dependent upon the retorting process used. In general, the higher the temperature to which the oil has been exposed, the higher the concentration of aromatics and the lower the concentration of saturates and olefins. In most cases, the simplest member of a homologous series is the predominant compound.

Upgrading, or partial refining, to improve the properties of a crude shale oil may be carried out using different options (Figures 4.4, 4.5, and 4.6), each with different objectives, depending on the composition of the crude shale oil and the intended use for the product.

Table 4.4: Properties of shale oil.

	ABOVE GROUND	IN SITU.	ARABIAN LIGHT
GRAVITY, °API	20.9	23.6	33.3
COMPOSITION, WT%			
CARBON	84.92	85.80	85.15
HYDROGEN	11.43	11.01	12.80
NITROGEN	1.97	1.42	0.09
OXYGEN	1.04	1.21	0.13
SULFUR	0.62	0.53	1.80
METALS, PPM	106	67	27
H/C ATOMIC RATIO	1.615	1.539	1.803

What is perhaps exceptional about the composition of crude shale oil (in addition to the hydrocarbon distribution; Figures 4.6 and 4.7) is the high nitrogen content, relative to the nitrogen content of crude petroleum (Table 4.4). As is the case with crude

128

Table 4.5: Properties of shale oil residua.

	ABOVE GROUND	IN SITU	ARABIAN LIGHT
GRAVITY, °API	10.35	10.50	9.97
COMPOSITION, WT%			
CARBON	84.32	84.29	85.99
HYDROGEN	10.60	11.09	10.12
NITROGEN	2.45	1.54	0.28
OXYGEN	2.05	2.51	0.25
SULFUR	0.58	0.49	3.32
METALS	182	176	100
H/C ATOMIC RATIO	1.508	1.577	1.412

petroleum, much of the nitrogen is carried over into the residua (Table 4.5). In terms of refining and catalyst activity, the nitrogen content of shale oil is a disadvantage. However, in terms of the use of shale oil residuum as a modifier for asphalt, where nitrogen species can enhance binding with the inorganic aggregate, the nitrogen content is beneficial. This is especially true, since much of the nitrogen occurs in aromatic systems and is basic in nature (Figure 4.8)

For example, the treating may be a cursory, partial, refining treatment to produce an oil that can be transported to a refinery by pipeline (i.e., meets the pipeline operator's specifications and also stability criteria). Alternatively, the treatment may involve more complete upgrading to produce a premium refinery feedstock with low-nitrogen, low-sulfur, and low-residuum feedstock to yield the finished products (e.g., gasoline, diesel fuel, and jet fuel) and which will also afford feedstocks for chemical/petrochemical operations.

If not removed, the arsenic and iron in shale oil would poison and foul the supported catalysts used in hydrotreating. Because these materials are soluble, they cannot be removed by filtration. Several methods have been used specifically to remove arsenic and iron. Other methods involve hydrotreating; these also lower sulfur, olefin, and diolefin contents and, thereby, make the upgraded product less prone to form gum.

Thermal conversions, coking, and visbreaking, are conceptually simple, noncatalytic methods for lowering the high pour point and viscosity of raw shale oils in order to make the oil more suitable for hydrotreating that is needed to remove nitrogen and sulfur. Coking also separates suspended solids.

Hydrotreating is more flexible and less destructive than coking as a way to remove nitrogen, sulfur, oxygen, arsenic, and metals. One approach has been to distill the crude shale oil, then to hydroprocess the fractions.

In addition to being rich in nitrogen species (such as alkylpyridines and/or quinolines, alkylpyrroles/indoles, and nonbasic alkylpyrroles/indoles/carbazoles/benzocarbazoles), shale oil is also rich in high-molecular-weight, waxy paraffinic material.

Thermal cracking lowers the molecular weight of shale oil feedstocks but yields straight-chain products of low octane number.

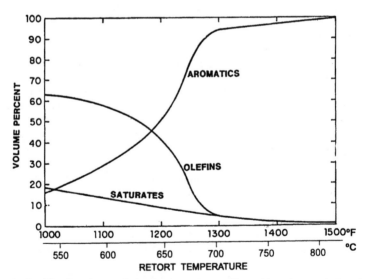

Figure 4.7: Variation of hydrocarobon distribution with time

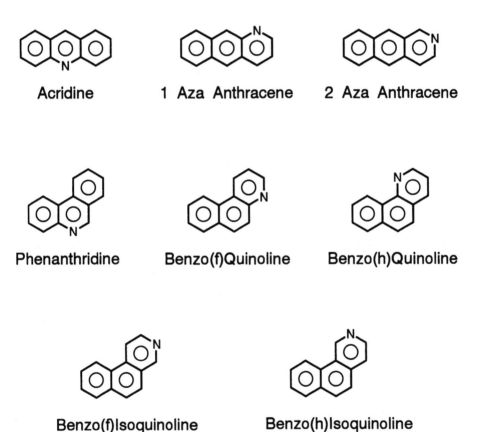

Acridine   1 Aza Anthracene   2 Aza Anthracene

Phenanthridine   Benzo(f)Quinoline   Benzo(h)Quinoline

Benzo(f)Isoquinoline   Benzo(h)Isoquinoline

Figure 4.8: Examples of three-ring nitrogen species which occur in shale oil

131

Fluid catalytic cracking not only lowers molecular weight but also causes isomerization to produce branched products with higher octane numbers. As a result, catalytically cracked shale oil naphtha is a more desirable feedstock to produce a gasoline blending stock (by hydrotreating) than is the naphtha from thermal cracking or from the coking of shale oil.

Hydroprocessing of the raw shale oil to remove basic nitrogen that would poison the acidic cracking catalyst, and would also promote unwanted coking, is required as a pretreatment to fluid bed catalytic cracking.

If shale oil liquids are to be used as liquid fuels the production (retorting) and process (refining) conditions and the quality requirements of the oil shale derived products must be optimized.

Severe hydroprocessing of the raw shale oil may be a satisfactory solution because of the anticipated high-stability properties of the products. Mild to intermediate-severity hydroprocessing followed by blending with conventional petroleum products could also be an acceptable compromise, although it might be anticipated that the shale oil would retain some of the instability-inducing constituents.

In addition to being a potential source of hydrocarbon products, oil shale can also be regarded as a source of other products, such as asphalt. It is such a product that could provide the "market pull," thereby circumventing the politics of oil (Chapter 1) and promoting the development of an oil shale industry. This would provide the incentive to develop domestic energy resources and be free from foreign energy-pricing policies.

Approximately 2 million miles, or 93% of all road surfaces in the United States are paved with asphalt, and approximately 21 million tons of asphalt and asphalt-related paving materials are used annually. Expenditures are in excess of $10 billion each year for the construction and maintenance of asphalt roadways.

In addition, the Road Information Program (TRIP) in Washington, D.C., has reported that "bouncing over rough and broken pavements is costing the U.S. motorist more than $28 billion annually in added vehicle operating cost. This total included $21.7 billion for wasted fuel and another $6.5 billion spent on repairs to tires, brakes, and steering and suspension systems."

In recent years there has been an increasing trend in the U.S.

refining industry to process the heavy end of the crude oil barrel into lighter distillate and transpiration fuels, at the expense of producing petroleum binders for asphalt. This recent and marked increase in "bottom of the barrel" processing has resulted in periodic shortages of domestically produced asphalt binders and a greater reliance on imported petroleum feedstocks for the production of asphalt binders.

There is increasing concern that asphalt produced from these new or untried crude blends is causing an increasing number of pavement problems, which are exacerbated by the significant increase in traffic frequency and increased truck weights in recent years. These factors are combining to create greater stresses on, and an increasing rate of failure of, asphalt roadways.

Moisture damage and binder embrittlement are major causes of pavement failure. Moisture damage results from a weakening or disruption of the asphalt-aggregate bond, leading to a loss of the structural strength of the pavement. This can produce pavement stress symptoms such as rutting, raveling, flushing, and cracking that may eventually lead to pavement breakup.

In addition to inadequacies in pavement design and construction, moisture damage typically is enhanced by such factors as deficiencies in the asphalt, moisture-sensitive aggregates, environmental conditions, and increased traffic loading.

On the other hand, binder embrittlement results when the flow properties of the asphalt binder deteriorate to such an extent that, under the influence of physical or thermal stresses, the binder, and thus the pavement, fractures.

The residuum from vacuum distillation of crude shale oil can produce materials with consistencies in the asphalt range with yields of the order of 10-20% depending on the retorting process that was used.

3.0 Stability of Oil Shale Products

Just as petroleum fuels are complex mixtures of various compounds, so is shale oil, but, perhaps even more so! Whilst the constituents can be generally categorized as hydrocarbons, nitrogen compounds, oxygen compounds, and sulfur compounds, there is more to shale oil composition than mere general categorization. There are also metallic species (particularly arsenic derivatives), and it is the form (Scouten, 1990) in which the constituents exist that plays a major role in instability/incompatibility.

There are several alternatives for stabilizing the constituents of shale oil, thereby preventing the adverse effects of continued chemical reaction. These methods are (see also Figure 4.9):

(1) hydroprocessing the shale oil at the retorting site;
(2) adding antioxidant to the freshly produced shale oil;
(3) blending the freshly obtained shale oil with a conventional crude oil in predetermined suitable proportions;
(4) prevention of exposure of the shale oil to air (oxygen) by using an inert gas blanket; and
(5) any suitable combination of the above steps.

Hydroprocessing the shale oil will yield a reasonably stable product. However, the products (gasoline, jet fuel, or diesel fuel) may require further hydroprocessing before becoming the final products.

Catalytic hydrogenation of a freshly produced shale oil yields a product that is fairly stable for limited periods. The severity of the hydrogenating process determines the stability level of the product (Table 4.6) (Por, 1992).

Changes in the various properties are evident after the shale oil has been hydroprocessed under various conditions of increasing

Table 4.6: Properties of hydrotreated shale oil.

Property	Raw shale oil	Hydrotreated shale oil
Gravity, °API	11	18-23
Nitrogen, wt%	1.4	1
Sulfur, wt%	5	1
Carbon residue		
Conradson	3	1
Asphaltenes	5	1

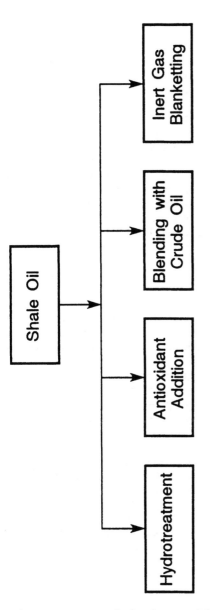

Figure 4.9: Methods for stabilizing and use of shale oil.

severity, for example, there are decreases in the densities, viscosity, sulfur content, acidity, bromine number, ash content, arsenic content, asphaltene content, bottom sediment, and water content.

Blending shale oil products with corresponding crude oil products, using shale oil fractions obtained from a very mildly hydrogen-treated shale oil, yields kerosene and diesel fuel of satisfactory stability properties. However, the stability is for limited periods only and only for blends containing relatively low proportions of the oil shale derived components (Por, 1992).

Hydroprocessing of such shale oil products, either alone or in a blend with the corresponding crude oil fractions, is therefore necessary. The severity of the hydroprocessing has to be adjusted according to the particular properties of the feed and the required level of the stabilities of the product.

The **degree of unsaturation** of motor gasolines obtained from treated shale oil has a significant effect on the stability properties of the product. It has been found that a content of up to 20% olefins, in the presence of 100 ppm of a tertiary butyl phenol antioxidant, reduces the stability properties of motor gasolines but still leaves them within the limits acceptable by most of the quality specifications.

Thus, a blend of such a shale oil gasoline with a straight-run gasoline, or a catalytically reformed gasoline in suitable ratios, should yield a product of satisfactory stability properties, provided that the aromatics content does not exceed certain limits. The synergistic effect of other constituents should also be taken into consideration.

The synergistic effect of diolefins, even if present in only small amounts, is significant in increasing the rate of degradation products formation. The presence of as little as 2% diolefins may increase the gum-forming tendency of a gasoline, containing 10% olefins, far above the normally acceptable values (Por, 1992).

Aromatic components, in concentrations of up to 40-50%, do not reduce the induction periods of motor gasolines appreciably, provided that the remaining 50-60% of the other components are themselves stable and contribute to the overall stability of the product. However, in long-term storage, even such concentrations of aromatic constituents might affect the stability properties of this product unfavorably.

The maximum concentration of aromatics in gasoline intended for long-term storage has been found to amount to 30-35%, provided that practically no olefins are present and that the total sulfur content does not exceed 0.05-0.10%, depending on the composition of the sulfur compounds. In conjunction with olefins and diolefins, the aromatic components may contribute to the instability of gasoline blends. A blend consisting of, say, 15% olefins and 85% saturated constituents would be more stable than a blend containing 15% olefins, 30% aromatics, and 55% saturated constituents.

Gasoline from shale oil usually contains a high percentage of aromatic and naphthenic compounds that are not affected by the various treatment processes. The olefin content, although reduced in most cases by refining processes, will still remain significant. It is assumed that diolefins and the higher unsaturated constituents will be removed from the gasoline product by appropriate treatment processes (Sullivan and Strangeland, 1979). The same should be true, although to a lesser extent, for nitrogen- and sulfur-containing constituents.

The sulfur content of raw shale oil gasoline may be rather high due to the high sulfur content of the shale oil itself and the frequently even distribution of the sulfur compounds in the various shale oil fractions. The concentration as well as the type of sulfur compounds are both of importance when studying their effect on the gum formation tendency of the gasolines containing them.

Sulfides (R-S-R), disulfides (R-S-S-R), and mercaptans (R-SH) are, among the other sulfur compounds, the major contributors to the gum formation in gasoline. Sweetening processes for converting mercaptans to disulfides should, therefore, not be used for shale oil gasoline; sulfur extraction processes (such as the Merox process) are preferred.

Catalytic hydrodesulfurization processes are not a good solution for the removal of sulfur constituents from gasoline when high proportions of unsaturated constituents are present. A significant amount of the hydrogen would be used for hydrogenation of the unsaturated components. However, when hydrogenation of the unsaturated hydrocarbons is desirable, catalytic hydrogenation processes would be effective.

Gasoline derived from shale oil contains varying amounts of oxygen compounds. The presence of oxygen in a product in which free radicals form easily, is a cause for concern. Free hydroxy radicals are generated, and the polymerization chain reaction is quickly brought to its propagation stage. Unless effective means are provided for the termination of the polymerization process, the propagation stage may well lead to an uncontrollable generation of oxygen-bearing free radicals, leading in turn to gum and other polymeric products.

Diesel fuel derived from oil shale is also subject to the degree of unsaturation, the effect of diolefins, the effect of aromatics, and the effect of nitrogen and sulfur compounds. In addition, the larger molecules involved in the case of the higher-molecular-weight diesel fuels make the situation more complex.

Jet fuel produced from shale oil would have to be subjected to suitable refining treatments and special processes. The resulting product must be identical in its properties to corresponding products obtained from conventional crude oil. This can be achieved by subjecting the shale oil product to a severe catalytic hydrogenation process with subsequent addition of additives to ensure resistance to oxidation.

If antioxidants are used for a temporary reduction of shale oil instability, they should be injected into the shale oil (or its products) as soon as possible after production of the shale oil. The antioxidant types and their concentrations should be determined separately for each particular case.

The antioxidants combine with the free radicals or supply available hydrogen atoms to mitigate the progress of the propagation and branching processes. When added to the freshly produced unstable product, the antioxidants may be able to fulfill this purpose. However, when added after some delay, i.e. after the propagation and the branching processes have advanced beyond controllable limits, the antioxidants would not be able to prevent formation of degradation products.

Exposure to oxygen is a major factor contributing to degradation product formation in shale oils. Peroxy radicals, which are readily formed when untreated shale oils or their products are exposed to oxygen, lead to a rapid gum formation rate. Once oxygen is eliminated from such a system, the polymerization chain reaction tends to arrive at its termination stage, a process that can take place in several ways, one way being by exhaustion of the reactive monomers or a combination of two free radicals. Chain reaction termination can also be affected by radical combination or disproportionation.

In all cases, free radicals have to be eliminated from the system. The chain termination can also be induced by certain constituents present naturally or added artificially in the form of antioxidants.

4.0 References

Allred, V.D. (editor). 1982. Oil Shale Processing Technology. Center for Professional Advancement, East Brunswick, New Jersey.
Durand, B. (editor). 1980. Kerogen: Insoluble Matter from Sedimentary Rocks. Editions Technip, Paris, France.
Hisamitsu, T., Gomyo, K., and Maruyama, F. 1993. Sekiyu Gakkaishi. 36: 485.
Maciel, G.E., V.J. Bartuska, and F.P. Miknis. 1978a. Fuel. 58:155.
Maciel, G.E., V.J. Bartuska, and F.P. Miknis. 1978b. Fuel. 57: 505.
Por, N. 1992. Stability Properties of Petroleum Products. Israel Institute of Petroleum and Energy, Tel Aviv, Israel.
Russell, P.L. (editor). 1980. History of Western Oil Shale. Center for Professional Advancement, East Brunswick, New Jersey.
Scouten, C. 1990. In Fuel Science and Technology Handbook. J.G. Speight (editor). Marcel Dekker Inc., New York.
Sinor, J.E. 1988. Proceedings. International Conference on Oil Shale and Shale Oil. Chemical Industry Press, Beijing, China. p. 159
Sullivan, R.F., and Strangeland, B.E. 1979. In Refining of Synthetic Crudes. M.L. Gorbaty and B.M. Harney (editors). Advances in Chemistry Series No. 179. American Chemical Society, Washington, DC. Chapter 3.
Yen, T.F. (Editor). 1976. Science and Technology of Oil Shale. Ann Arbor Science Publishers Inc., Ann Arbor, Michigan.

CHAPTER 5: LIQUID FUELS AND OTHER PRODUCTS

1.0 Introduction

Petroleum products had been in use for over five thousand years, when it was recognized that the heavier derivatives of petroleum (the nonvolatile residua and asphalt) could be used for caulking and as an adhesive for jewelry or for construction purposes. There is also documented use of asphalt for medicinal purposes (Table 5.1) (Abraham, 1945; Forbes, 1958; Speight, 1991).

The petroleum industry has been inspired by the technological advances of the twentieth century. Or perhaps the converse of this statement is true!

However, for the main part, the petroleum industry was inspired by the development of the automobile and the continued demand for gasoline and other fuels (Hoffman, 1992). Such a demand has been accompanied by the demand for other products: diesel fuel for engines, lubricants for engine and machinery parts, fuel oil to provide power for the industrial complex, and asphalt for roadways.

There are a myriad of other products that have evolved through the short life of the petroleum industry, and the complexities of product composition have matched the evolution of the products (Hoffman, 1992). In fact, it is the complexity of product composition that has served the industry well and, at the same time, had an adverse effect on product use. Product complexity has made the industry unique among industries. Indeed, current analytical techniques that are accepted as standard methods for, as an example, the aromatics content of fuels (ASTM D 1319, ASTM D 2425, ASTM D 2549, ASTM D 2786, ASTM D 2789; American Society for Testing and Materials, 1993), as well as proton and carbon nulcear magnetic resonance methods, yield different information. Each method will yield the "% aromatics" in the sample, but the data must be evaluated within the context of the method.

Product complexity, and the means by which the product is evaluated, has made the industry unique among industries. But product complexity has also brought to the fore issues such as instability and incompatibility. In order to understand the evolution of the products it is essential to have an understanding of the composition of the various products.

Product complexity becomes even more meaningful when various fractions from different types of crude oil as well as fractions

Table 5.1:   Petroleum and derivatives such as asphalt have
              been known and used for almost six thousand years.

3800 BC	First documented use of asphalt for caulking reed boats.
3500 BC	Asphalt used as cement for jewelry and for ornamental applications.
3000 BC	Use of asphalt as a construction cement by Sumerians; also believed to be used as a road material; asphalt used to seal bathing pool or water tank at Mohenjo Daro.
2500 BC	Use of asphalt and other petroleum liquids (oils) in the embalming process; asphalt believed to be widely used for caulking boats.
1500 BC	Use of asphalt for medicinal purposes and (when mixed with beer) as a sedative for the stomach; continued reference to use of asphalt liquids (oil) as illuminant in lamps.
1000 BC	Use of asphalt as a waterproofing agent by lake dwellers in Switzerland.
500 BC	Use of asphalt mixed with sulfur as an incendiary device in Greek wars; also use of asphalt liquid (oil) in warfare.
350 BC	Occurrence of flammable oils in wells in Persia.
300 BC	Use of asphalt and liquid asphalt as incendiary device (Greek fire) in warfare.
250 BC	Occurrences of asphalt and oil seepages in several areas of the "fertile crescent" (Mesopotamia); repeated use of liquid asphalt (oil) as an illuminant in lamps.
750 AD	Use in Italy of asphalt as a color in paintings.
950 AD	Report of destructive distillation of asphalt to produce an oil; reference to oil as nafta (naphtha).
1500 AD	"Discovery" of asphalt deposits in the Americas; first attempted documentation of the relationship of asphalts and naphtha (petroleum).
1600 AD	Asphalt used for a variety of tasks; relationship of asphalt to coal and wood tar studied; asphalt studied; used for paving; continued documentation of the use of naphtha as an illuminant and the production of naphtha from asphalt; importance of naphtha as fuel realized.
1859 AD	Discovery of petroleum in North America; birth of modern-day petroleum science and refining.

from synthetic crude oil are blended with the corresponding petroleum stock. The implications for refining the fractions to salable products increase (Dooley et al., 1979).

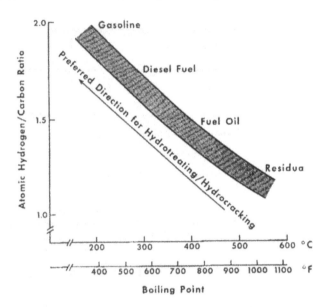

Figure 5.1: Relative boiling ranges of petroleum products.

Synthetic crude oil (syncrude) is a complex mixture of hydrocarbons, somewhat similar to petroleum, obtained from the bituminous material that is found in tar sand formations, from coal (liquefaction), from synthesis gas (a mixture of carbon monoxide and hydrogen), or from oil shale. Syncrudes generally differ in composition from petroleum; for example, syncrude from coal usually contains more aromatic hydrocarbons than petroleum, and this is often the cause of incompatibility.

Briefly, petroleum refining is the separation of petroleum into fractions and the subsequent treating of these fractions to yield marketable products (Chapter 2).

Most liquid fuels and other products are either fractions of petroleum that have been treated to remove undesirable components

143

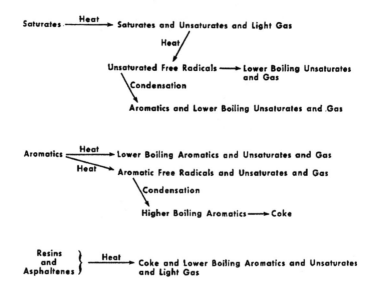

Figure 5.2: Representation of thermal decomposition pathways.

or are totally, or partly, synthetic, in that they have compositions that are impossible to achieve by direct separation of these materials from crude petroleum. They result from chemical processes that change the molecular nature of selected portions of crude petroleum. In addition, there is a general increase in aliphatic character (increase in H/C ratio) as the molecular weight of the product is decreased from that of a residuum to a liquid fuel (Figure 5.1).

Part of the problem that arises during refining, and that can lead to instability/incompatibility, is the complex chemistry that is a result of the refining conditions. Whatever the indigent fraction, unsaturated products are produced (Figure 5.2), and there is no guarantee of "sameness." Rather, it might be expected that the products from one fraction will be quite dissimilar from the analogous products from another fraction.

For example, and in terms of instability/incompatibility,

phenolic compounds have an adverse effect on the storage stability of fuel oil and, by inference, diesel fuel and other fuels (Green et al., 1992). By inference, the addition of coal liquids (containing phenols) to hydrocarbon fuels could have a severely adverse effect on the properties of the fuel. The option is some degree of hydrotreating to remove the phenols.

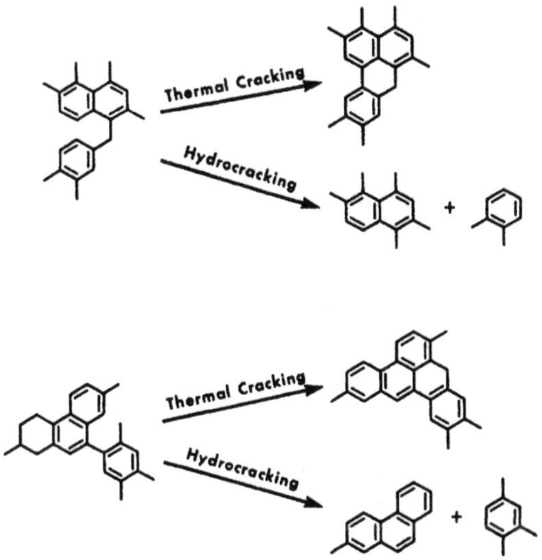

Figure 5.3: Illustration of coking and hydrocracking chemistry.

In addition, there is the general belief that the formation of coke involves immediate condensation of aromatic species that is prevented when hydrogen is present (Figure 5.3). This may not be the case (Chapter 13), since it is now recognized that the initial steps in the formation of coke probably involve the production of lower-molecular-weight polar species. Thus, chemical formulae might be valuable in describing some of the reactions that occur, but they can also be misleading.

Coal (Chapter 3) and oil shale (Chapter 4) may also be used as sources of liquid fuels but introduce the aspect of different compositions by virtue of the different nature of the source materials. Blending such streams with crude oil prior to refining or with petroleum products can introduce a situation that can result in instability or incompatibility of the blended materials.

Therefore, in order to understand the phenomena of instability and incompatibility it is necessary to first understand the nature and behavior of the products as evidenced by their composition.

It is the goal of this chapter to illustrate the nature of the products derived from various sources (predominantly petroleum) and to examine the composition of the products and how such composition might influence stability and/or incompatibility.

2.0 Gasoline

Gasoline is a mixture of hydrocarbons that boil below 200°C (390°F). The hydrocarbon constituents in this boiling range are those that have four to 12 carbon atoms in their molecular structure (Owen, 1973). Thus, gasoline can vary widely in composition; even gasolines with the same octane number may have different compositions, not only in terms of the proportion of the constituents, but also in terms of having different constituents.

For example, low-boiling distillates with high aromatics contents (above 20%) can be obtained from some crude oils. The variation in aromatics content as well as the variation in the content of normal paraffins, branched paraffins, cyclopentanes, and cyclohexanes all involve characteristics of any one individual crude oil and influence the octane number of the gasoline.

Up to, and during, the first decade of the twentieth century, the gasoline produced from crude oil was that which was originally present in crude oil or which could be condensed from natural gas. However, after the discovery that the less volatile (and even the nonvolatile) constituents of petroleum (such as the fractions that boiled higher than kerosene, e.g., gas oil and the residuum) were heated to more severe temperatures, thermal degradation (or cracking) occurred to produce lower-molecular-weight products that were within the range suitable for gasoline (Chapter 2).

At first, gasoline produced by cracking was regarded as an inferior product because of its comparative instability on storage. As the demand for gasoline increased and the inability of refiners to produce **straight-run gasoline** decreased, the petroleum industry revolved around processes by which gasoline could be produced (e.g.

catalytic cracking, thermal and catalytic reforming, hydrocracking, alkylation, and polymerization; Chapter 2), and the problem of storage instability was addressed with some success.

In addition, the differences in composition of various gasoline products initiated the concept of gasoline blending to produce an **average** product.

The physical process of blending the gasoline components(Table 5.2) is relatively simple, but determination of how much of each component to include in a blend is much more difficult.

The operation is carried out by simultaneously pumping all the components of a gasoline blend into a pipeline that leads to the gasoline storage. However, the pumps must be set to automatically deliver the proper proportion of each component. Baffles in the

Table 5.2:   Component streams for gasoline.

Stream	Producing Process	Boiling Range °C	°F
Paraffinic			
Butane	Distillation Conversion	0	32
Isopentane	Distillation Conversion Isomerization	27	81
Alkylate	Alkylation	40-150	105-300
Isomerate	Isomerization	40- 70	105-160
Straight-run naphtha	Distillation	30-100	85-212
Hydrocrackate	Hydrocracking	40-200	105-390
Olefinic			
Catalytic naphtha	Catalytic Cracking	40-200	105-390
Steam cracked naphtha	Steam Cracking	40-200	105-390
Polymer	Polymerization	60-200	140-390
Aromatic			
Catalytic reformate	Catalytic Reforming	40-200	105-390

pipeline are often used to mix the components as they travel to the storage tank.

Aviation gasoline, usually found in use in light aircraft and older civil aircraft, has a narrower boiling range than conventional (automobile) gasoline. Thus, a boiling range of 38-170°C (100-340°F), compared to -1-200°C (30-390°F) for automobile gasolines, is realized for aviation gasoline.

The narrower boiling range ensures better distribution of the vaporized fuel through the more complicated induction systems of aircraft engines. Since aircraft operate at altitudes where the prevailing pressure is less than the pressure at the surface of the earth (pressure at 17,500 ft is 7.5 psi (0.5 atmosphere) compared with 14.8 psi (1.0 atmosphere) at the surface of the earth), the vapor pressure of aviation gasolines must be limited to reduce boiling in the tanks, fuel lines, and carburetors.

In contrast to the gasoline fraction produced from petroleum, shale oil gasoline contains considerably more nitrogen and is somewhat more acidic (Scouten, 1990). The nitrogen content can generally be reduced by hydrotreatment.

3.0 Naphtha

Petroleum naphtha has been available since the early days of the petroleum industry. Indeed, the infamous **Greek fire** documented as being used in warfare during the last three millennia is a petroleum derivative. It was produced either by distillation of crude oil isolated from a surface seepage or (more likely) by destructive distillation of the bituminous material obtained from tar pits (of which there are/were many known during the heyday of the civilizations of the Fertile Crescent). The bitumen obtained from the area of Hit (Is) in Iraq (Mesopotamia) is an example of such an occurrence (Abraham, 1945; Forbes, 1958; Speight, 1991).

Naphtha is a generic term that is applied to refined, partly refined, or unrefined petroleum products (Boenheim and Pearson, 1973; Hadley and Turner, 1973). It is valuable as a solvent because of the good dissolving power. The wide range of naphtha available from crude oil refining and the varying degree of volatility possible offer products suitable for many uses.

Naphtha is prepared by any one of several methods including fractionation of distillates or even crude petroleum, solvent extraction, hydrogenation of distillates, polymerization of unsaturated (olefinic) compounds, and alkylation processes. The naphtha may also be a combination of product streams from more than one of these processes.

The main uses of petroleum naphtha fall into the general areas of solvents (diluents) for paints, for dry cleaning, for cutback asphalt, in the rubber industry, and for industrial extraction processes. Turpentine, the older, more conventional solvent for paints, has now been almost completely replaced by the cheaper and more abundant petroleum naphtha.

4.0 Kerosene

Kerosene was the major refinery product before the onset of the "automobile age," but now kerosene might be termed as one of several other petroleum products after gasoline.

Kerosene originated as a straight-run (distilled) petroleum fraction that boiled between approximately 205 and 260°C (400-500°F). In the early days of petroleum refining some crude oils contained kerosene fractions of very high quality, but other crude oils, such as those having a high proportion of asphaltic materials, must be thoroughly refined to remove aromatics and sulfur compounds before a satisfactory kerosene fraction can be obtained.

The kerosene fraction is essentially a distillation fraction of petroleum (Walmsley, 1973). The quantity and quality of the kerosene vary with the type of crude oil; some crude oils yield excellent kerosene, but others produce kerosene that requires substantial refining. Kerosene is a very stable product, and additives are not required to improve the quality. Apart from the removal of excessive quantities of aromatics, kerosene fractions may need only a lye (alkali) wash if hydrogen sulfide is present.

The kerosene fraction from shale oil (Por, 1992) is like the gasoline/naphtha fraction, generally high in nitrogen and acidity. However, hydrotreatment will remove most of the nitrogen, but catalyst degeneration can be quite severe.

5.0 Fuel Oil

Fuel oil is classified in several ways but generally may be divided into two main types: **distillate fuel oil** and **residual fuel oil** (Pope, 1973).

Distillate fuel oil is vaporized and condensed during a

149

distillation process. Therefore, distillate fuel oil has a definite boiling range and does not contain high-boiling oils or asphaltic components.

However, for the purposes of this chapter, it is worth defining the different fuel oils to alleviate any potential confusion that may arise from the terminology.

No. 1 fuel oil is very similar to kerosene and is used in burners where vaporization before burning is usually required and a clean flame is specified. **No. 2 fuel oil** is often called **domestic heating oil** and has properties similar to diesel fuel and heavy jet fuel; it is used in burners where complete vaporization is not required before burning. **No. 4 fuel oil** is a light industrial heating oil and is used where preheating is not required for handling or burning; there are two grades of No. 4 fuel oil, differing in safety (flash point) and flow (viscosity) properties. **No. 5 fuel oil** is a heavy industrial fuel oil that requires preheating before burning. **No. 6 fuel oil** is also a heavy fuel oil and is more commonly known as **Bunker C oil** when it is used to fuel ocean-going vessels; preheating is always required for burning this oil.

A fuel oil that contains any amount of the residue from crude distillation or thermal cracking is a residual fuel oil. The terms distillate fuel oil and residual fuel oil are losing their significance, since fuel oil is now made for specific uses and may be either distillates, residuals, or mixtures of the two. The terms **diesel fuel, domestic fuel oil**, and **heavy fuel oil** are more indicative of the uses of fuel oil.

Diesel fuel is somewhat better defined compared with the other fuel oils. The specifications indicate the maximum levels of sulfur as well as minimal physical properties that must be met to ensure fuel reliability. Boiling range is also a necessary specification for diesel fuel.

Similarly, jet fuel, which can be considered one of the fuel oils, is also subject to rigid specifications.

Domestic fuel oil is fuel oil used primarily in the home and includes kerosene, stove oil, and furnace fuel oil. The use of diesel fuel oil is self-explanatory, but residual oil has been successfully used to power marine diesel engines, and mixtures of distillate fuel oil and residual fuel oil have been used on locomotive diesels.

Heavy fuel oil includes a variety of components ranging from distillate to residual oil that must be heated to 260þC (500þF) or higher before they can be used. In general, heavy fuel oil consists of residual oil blended with distillate to suit specific needs. Included in heavy fuel oil are various industrial oils, and when used to fuel ships, heavy fuel oil is called **bunker oil**.

Heavy fuel oil usually contains residuum, which is mixed (cut back) to a specified viscosity with gas oil and fractionator residua. For some industrial purposes where flames or flue gases contact the product (ceramics, glass, heat treating, open hearth furnaces), fuel oil must be blended to contain minimum sulfur contents; low-sulfur residua are preferable for these fuels.

Stove oil is a straight-run (distilled) fraction from crude oil, whereas other fuel oil is usually a blend of two or more fractions. The straight-run fractions available for blending into fuel oil are heavy naphtha, light and heavy gas oil, and residua. Cracked fractions such as light and heavy gas oil from catalytic cracking, cracking coal tar, and fractionator residua from catalytic cracking may also be used as blends to meet the specifications of different fuel oils.

The manufacture of fuel oil at one time largely involved using what was left after removing desired products from crude petroleum. Now fuel oil manufacture is a complex matter of selecting and

Table 5.3: Viscosity of various lubricating oil grades.

| | Kinematic viscosity, cs @ 100°C (212°F) | | Dynamic viscosity, cp |
	minimum	maximum	maximum
SAE 10W	4.1		3500 @ -20°C (-4°F)
SAE 15W	5.6		3500 @ -15°C (5°F)
SAE 20W	5.6		4500 @ -10°C (14°F)
SAE 25W	9.3		6000 @ -5°C
SAE 20	5.6	9.3	
SAE 30	9.3	12.5	
SAE 40	12.5	16.3	
SAE 50	16.3	21.9	

blending various petroleum fractions to meet definite specifications.

6.0 Lubricating Oil

Lubricants, either as oil (Mills, 1973) or semisolid grease (Dawtrey, 1973), have been known about for centuries (if not millennia), and prior to the onset of petroleum refining the preferred lubricants were materials such as lard oil, whale oil, and tallow.

The same demand that existed for kerosene did not exist for petroleum-derived lubricating oil; nor is it clear that the early refiners knew of the value of petroleum as a source of lubricating stocks. After kerosene, early petroleum-refining operations focused on the production of paraffin wax for the manufacture of candles. Lubricating oil was, at first, a by-product of wax manufacture.

However, as the need for lubricants increased (in parallel with the evolution of the automobile engine and other moving machinery parts) and the trend to heavier industry increased, the demand for lubricating oil increased. With the discovery that petroleum could supplement the increased demand for lubricants, after the 1890s, petroleum largely replaced animal oil and vegetable oil as the source of lubricants. However, over the past five decades, synthetic lubricants have emerged as a major product (Gunderson and Hart, 1962).

Many of the synthetic lubricants are of a different chemical nature from those that are petroleum based, thereby increasing the possibility of incompatibility when petroleum-based and synthetic lubricants are blended.

Lubricating oil is distinguished from other fractions of crude oil by a high (>400°C; >750°F) boiling point, as well as a high viscosity. In fact, lubricating oil is classified on the basis of viscosity (Table 5.3). This classification is based on the SAE (Society of Automotive Engineers) J 300 specification. The single-grade oils (e.g., SAE 20) correspond to a single class and have to be selected according to engine manufacturers' specifications, operating conditions, and climatic conditions. Multigrade lubricating oil such as SAE 10W-30 possesses at -20°C the viscosity of a 10W oil and at 100°C the viscosity of an SAE 30 oil.

Materials suitable for the production of lubricating oil are

152

comprised principally of hydrocarbons containing from 25 to 35 carbon atoms per molecule, whereas residua may contain molecular systems with 50-80 (or more) carbon atoms per molecule. Moreover, the molecular systems in residua are not usually suitable as constituents of lubricants.

The early method of lubricating oil manufacture (established during the 1870s and 1880s) depended on whether the crude petroleum was to be processed for kerosene, or for lubricating oil. Usually the crude oil was processed for kerosene and primary distillation separated the crude into three fractions: naphtha, kerosene, and a residuum. To increase the production of kerosene, the cracking distillation technique was used, and this converted a large part of the gas oil constituents and lubricating oil constituents into kerosene.

The development of vacuum distillation (Chapter 2) provided the means of separating more suitable lubricating oil fractions with predetermined viscosity ranges and removed the limit on the maximum viscosity that might be obtained in a distillate oil. Vacuum distillation prevented residua constituents (asphaltic material) from contaminating lubricating oil but did not remove other undesirable materials such as acidic components or components that caused the oil to thicken excessively when cold and become very thin when hot.

In general, lubricating oil may be divided into several categories according to the types of service they are intended to perform. More specifically, there are two main groups: (1) lubricating oil used in intermittent service, such as motor engine oil and aviation engine oil; and (2) lubricating oil designed for continuous service, such as turbine oil.

Lubricating oil used in intermittent service must show the least possible change in viscosity with temperature and must be changed at frequent intervals to remove the foreign matter that is collected during service. The stability of such lubricating oil, although important, is of somewhat lesser importance compared with the stability of lubricating oil used in continuous service for prolonged periods and without renewal. Lubricating oil used in continuous service must be extremely stable because the engines in which they are used operate at fairly constant temperature without frequent shutdown.

153

7.0 Wax

Wax is of two general types: the paraffin wax that occurs in distillates and the microcrystalline wax that occurs in residua (Mazee, 1973). The melting point of wax is not directly related to its boiling point because waxes contain hydrocarbons of different chemical structure. Nevertheless, waxes are graded according to their melting point and oil content.

Paraffin wax is a solid crystalline mixture of straight-chain (normal) hydrocarbons ranging from C_{20} to C_{30}, and higher. Wax constituents are solid at ordinary temperatures (25°C; 77°F), whereas **petrolatum** (petroleum jelly) contains both solid and liquid hydrocarbons.

Wax production by "wax sweating" was originally used to separate wax fractions with various melting points from the wax obtained from shale oil. Wax sweating is still used to some extent but is being replaced by the more convenient wax recrystallization process. In wax sweating, a cake of **slack wax** is slowly warmed to a temperature at which the oil in the wax and the lower-melting wax constituents become fluid and drip (or sweat) from the bottom of the cake, leaving a residue of higher-melting wax.

The amount of oil separated by sweating is now much smaller than it used to be due to the development of highly efficient solvent dewaxing techniques.

However, wax sweating may be considered to be a process that hinges on the incompatibility of the solid wax in the liquid constituents. The first liquid removed from the sweating pan is called **foots oil**, which melts at 38°C (100°F) or lower, followed by **interfoots oil**, which melts in the range 38-44°C (100-112°F). Crude scale wax next drips from the wax cake and consists of wax fractions with melting points above 44°C (112°F).

Wax recrystallization, like wax sweating, separates wax into fractions, but instead of relying upon differences in melting points, the process makes use of the different solubilities of the wax fractions in a solvent such as a ketone.

When a mixture of ketone and wax is heated, the wax usually dissolves completely, and if the solution is cooled slowly, a temperature is reached at which a crop of wax crystals is formed (solvent dewaxing; Figure 5.4). These crystals will all be of the same melting point, and if they are removed by filtration, a wax fraction with a specific melting point is obtained. If the clear

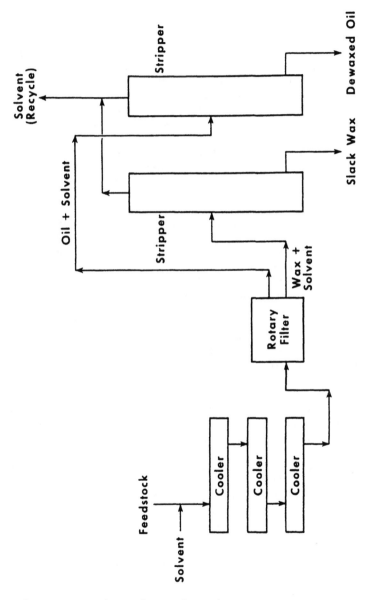

Figure 5.4: The solvent dewaxing process.

filtrate is cooled further, a second batch of wax crystals with a lower melting point is obtained. Thus, by alternate cooling and filtration, the wax can be subdivided into a large number of wax fractions, each with different melting points.

This method of producing wax fractions is much faster and more convenient than sweating and results in a much more complete separation of the various fractions. Recrystallization can also be applied to the purification of microcrystalline wax obtained from intermediate and heavy paraffin distillates, which cannot be sweated. The microcrystalline wax has a higher melting point than wax in general and differs in properties from paraffin wax.

8.0 Asphalt

Asphalt is a product of many petroleum refineries and may be residual asphalt, which is made up of the nonvolatile hydrocarbons in the feedstock, or may be produced by air blowing an asphaltic residuum (Figure 5.5). In addition, asphalt has a complex chemical and physical composition which usually varies with the source of the crude oil (Traxler, 1961; Barth, 1962; Hoiberg, 1964; Broome, 1973; Broome and Wadelin, 1973).

Asphalt manufacture is, in essence, a matter of distilling everything possible from crude petroleum until a residue with the desired properties is obtained. This is usually done by stages; distillation at atmospheric pressure removes the lower-boiling fractions and yields a reduced crude that may contain higher-boiling (lubricating) oil, asphalt, and even wax. Distillation of the reduced crude under vacuum removes the gas oil (and wax) as volatile overhead products, and the asphalt remains as a bottom (or residual) product. At this stage the asphalt is frequently (and incorrectly) referred to as pitch.

There are wide variations in refinery operations and in the type of crude oil; thus, different asphalt will be produced from different crude oil feedstocks. If lubricating oil is not required, the residuum may be distilled in a flash drum, which is similar to a distillation tower (but has few, if any, trays). Asphalt descends to the base of the flasher as the volatile components pass out of the top.

Asphalt is also produced by propane deasphalting (Figure 5.6). Asphalt can be made softer by blending the hard asphalt with the extract obtained in the solvent treatment of lubricating oil. On

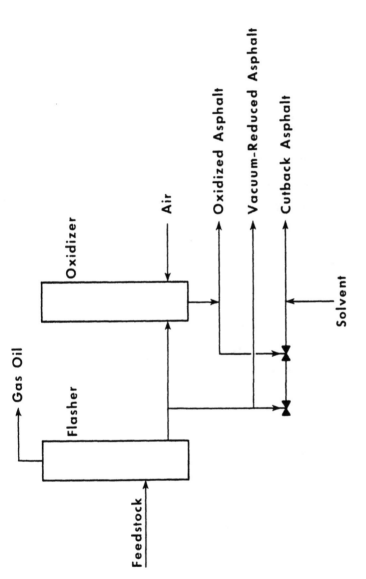

Figure 5.5: Asphalt manufacture.

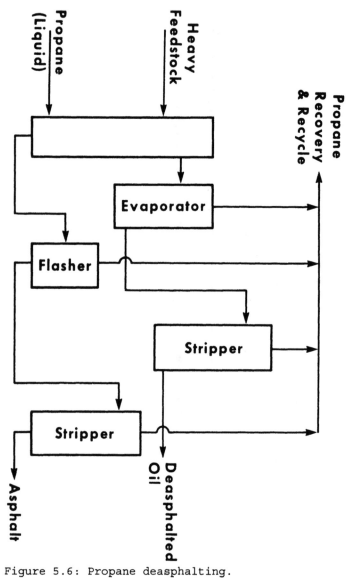

Figure 5.6: Propane deasphalting.

the other hand, soft asphalt can be converted into harder asphalt by oxidation (air blowing).

Road oil is a liquid asphalt material intended for easy application to earthen roads. They provide a strong base or a hard surface and will maintain a satisfactory passage for light traffic. Liquid road oil, cutback asphalt, and asphalt emulsion are of recent origin, but use of asphaltic solids for paving goes back to the European practices of the early 1800s.

Cutback asphalt is a mixture in which hard asphalt has been diluted with a lighter oil to permit application as a liquid without drastic heating. Cutback asphalt is classified as rapid, medium, and slow curing, depending on the volatility of the diluent, which also governs the rate of evaporation and consequent hardening.

An asphaltic material may be emulsified with water to permit application without heating. Such emulsions are normally of the oil-in-water type. They reverse or break on application to a stone or earth surface, so that the oil clings to the stone and the water disappears. In addition to their usefulness in road and soil stabilization, they are useful for paper impregnation and waterproofing. The emulsions are chiefly the soap or alkaline type and the neutral or clay type. The former break readily on contact, but the latter are more stable and probably lose water mainly by evaporation. Good emulsions must be stable during storage or freezing, suitably fluid, and amenable to control for speed of breaking.

As already pointed out (Table 5.1), asphalt has been known and used for six millennia or so. Nevertheless, it is only in the twentieth century that asphalt has grown to be a valuable refinery product. In the post-1980 period, a shortage of good-quality asphalt has developed in the United States. This is due in no short measure to the tendency of refineries in the post-1973 era to produce as much liquid fuels (e.g., gasoline) as possible. Thus, residua that would have once been used for asphalt manufacture are now being used to produce liquid fuels (and coke).

9.0 Coke

Petroleum coke is the residue left by the destructive distillation of petroleum residua. The coke formed in catalytic cracking operations is usually nonrecoverable because of adherence

159

to the catalyst, as it is often employed as fuel for the process.

The composition of petroleum coke varies with the source of the crude oil, but in general, large amounts of high-molecular-weight complex hydrocarbons (rich in carbon but correspondingly poor in hydrogen) make up a high proportion. The solubility of petroleum "coke" in carbon disulfide has been reported to be as high as 50-80%, but this is, in fact, a misnomer. Coke, by definition, is an insoluble, honeycomb-type material that is the end-product of thermal processes (Speight, 1991).

Coke is employed for a number of purposes, but the major use is in the manufacture of carbon electrodes for aluminum refining, which requires a high-purity carbon (low in ash and sulfur free). In addition, petroleum coke is employed in the manufacture of carbon brushes, silicon carbide abrasive, and structural carbon (pipes, Rashig rings, etc.), as well as in calcium carbide manufacture, from which acetylene is produced.

In the early days of refining (pre-1920s), coke was an unwanted refinery by-product and was usually discarded. As more coking operations became integral parts of refinery operations in the post-1920 era, coke was produced in significant amounts by many refiners. The demand for petroleum coke as an industrial fuel and for graphite/carbon electrode manufacture increased after World War II.

The use of coke as a fuel must proceed with some caution, with the acceptance by refiners of the heavier crude oils as refinery feedstocks. The higher content of sulfur and nitrogen in heavier crude oil means a coke product containing substantial amounts of sulfur and nitrogen. Both of these elements will produce unacceptable pollutants (sulfur oxides and nitrogen oxides) during combustion. These elements must also be regarded with caution in any coke that is scheduled for electrode manufacture, and removal procedures for these elements are continually being developed.

10.0 References

Abraham, H. 1945. Asphalt and Allied Substances. Volume I. 5th
Edition. Van Nostrand Inc., New York. P. 1.
American Society for Testing and Materials. 1993.   Annual Book
of Standards. Philadelphia, Pennsylvania.
Barth, E.J. 1962. Asphalt: Science and Technology. Gordon and
Breach, New York.
Boenheim, A.F., and Pearson, A.J. 1973. In Modern Petroleum
Technology. G.D. Hobson and W. Pohl (editors). Applied Science
Publishers Inc., Barking, Essex, England. Chapter 19.
Broome, D.C. 1973. In Modern Petroleum Technology. G.D. Hobson and
W. Pohl (editors). Applied Science Publishers Inc., Barking,
Essex, England. Chapter 23.
Broome, D.C., and Wadelin, F.A. 1973. In Criteria for Quality of
Petroleum Products. J.P. Allinson (editor). Halsted Press,
Toronto. Chapter 13.
Dawtrey, S. 1973. In Modern Petroleum Technology. G.D. Hobson and
W. Pohl (editors). Applied Science Publishers Inc., Barking,
Essex, England. Chapter 21.
Dooley, J.E., Lanning, W.C., and Thompson, C.J. 1979. In Refining
of Synthetic Crudes. M.L. Gorbaty and B.M. Harney (editors).
Advances in Chemistry Series No. 179. American Chemical
Society, Washington, DC. Chapter 1.
Forbes, R.J. 1958. A History of Technology. Volume V. Oxford
University Press, Oxford, England. P. 102.
Green, J.B., Stirling, K.Q., and Ripley, D.L. 1992. Proceedings.
4th International Conference on the Stability and Handling of
Liquid Fuels.   Report DOE/CONF-911102. U.S. Department of
Energy), Washington, DC. P. 503.
Gunderson, R.C., and Hart, A.W. (editors). 1962. Synthetic
Lubricants. Reinhold Publishing Corp., New York.
Hadley, D.J., and Turner, L. 1973. In Modern Petroleum Technology.
G.D. Hobson and W. Pohl (editors). Applied Science Publishers
Inc., Barking, Essex, England. Chapter 12.
Hoffman, H.L. 1992. In Petroleum Processing Handbook. J.J. McKetta
(editor). Marcel Dekker Inc., New York. P. 2.
Hoiberg, A.J. 1964. Bituminous Materials: Asphalts, Tar, and
Pitches. Interscience Publishers, New York.
Mazee, W.M. 1973. In Modern Petroleum Technology. G.D. Hobson and

W. Pohl (editors). Applied Science Publishers Inc., Barking, Essex, England. Chapter 22.

Mills, A.L. 1973. In Modern Petroleum Technology. G.D. Hobson and W. Pohl (editors). Applied Science Publishers Inc., Barking, Essex, England. Chapter 20.

Owen, K. 1973. In Modern Petroleum Technology. G.D. Hobson and W. Pohl (editors). Applied Science Publishers Inc., Barking, Essex, England. Chapter 15.

Pope, J.G.C. 1973. In Modern Petroleum Technology. G.D. Hobson and W. Pohl (editors). Applied Science Publishers Inc., Barking, Essex, England. Chapter 18.

Por, N. 1992. Stability Properties of Petroleum Products. Israel Institute of Petroleum and Energy, Tel Aviv, Israel.

Speight, J.G. 1991. The Chemistry and Technology of Petroleum. 2nd Edition. Marcel Dekker Inc., New York.

Traxler, R.N. 1961. Asphalt: Its Composition, Properties, and Uses. Reinhold Publishing Corp., New York.

Walmsley, A.G. 1973. In Modern Petroleum Technology. G.D. Hobson and W. Pohl (editors). Applied Science Publishers Inc., Barking, Essex, England. Chapter 17.

CHAPTER 6: CHEMICAL ASPECTS OF INCOMPATIBILITY

1.0 Introduction

Hydrocarbon stability varies with both the type of chemical bond and structure. Alkanes are the most stable compounds, with stability decreasing in the following order: alkanes >> branched alkanes >> monocyclics > polycyclics > alkenes, with the di- and tri-alkenes being the most reactive of the hydrocarbons. Thus, the majority of any deposit formed as a result of instability or incompatibility would consist of carbon from the reactive hydrocarbon species present.

Heteroatoms, particularly nitrogen, sulfur, and trace metals (Tables 6.1, 6.2, and 6.3), are present in petroleum and might also be expected to be present in liquid fuels and other products from petroleum. Indeed, this is often the case, although there may have been some skeletal changes induced by the refining process(es). Oxygen is much more difficult to define in petroleum and liquid fuels.

However, it must be stressed that instability/incompatibility is not directly related to the total nitrogen, oxygen, or sulfur content. The formation of color/sludge/sediment is a result of several factors. Perhaps the main factor is the location and nature of the heteroatom, which in turn, determines reactivity.

The origin of oxygen species in fuels is often suspect insofar as there is often conflicting evidence as to whether the oxygen is indigenous to the petroleum or whether it is "picked up" during processing. Be that as it may, oxygen occurs in petroleum and its products in hydroxyl (-OH), carboxyl (-COOH), ketone (>C=O), and ester (>COOR, where R is an alkyl or aromatic moiety) functions. Thus, debates about the precise origin of oxygen functions in petroleum may bear little relevance to incompatibility phenomena. It is essential to recognize that oxygen is present, and it must be dealt with accordingly.

Heteroatoms, oxygen, nitrogen, sulfur, and mineral matter have been found to comprise up to 40% of deposits that form on hot surfaces, with a correspondingly smaller percentage under ambient storage conditions (Nixon, 1962). The sulfur content of these deposits has been found to vary from 0.3% to 9% (Coordinating Research Council, 1979). Organic sulfur, 0.4% military and 0.3% commercial, is the most abundant heteroatom present in jet fuels.

Table 6.1 Representative Nitrogen-Containing Compounds in
Crude Oils

Nonbasic		
Pyrrole	C_4H_5N	
Indole	C_8H_7N	
Carbazole	$C_{12}H_9N$	
Benzo (α) carbazole	$C_{16}H_{11}N$	
Basic		
Pyridine	C_5H_5N	
Quinoline	C_9H_7N	
Indoline	C_8H_9N	
Benzo (α) quinoline	$C_{13}H_9N$	

Trace levels of certain sulfur compounds have been found to
also influence the deposit formation process under ambient
conditions. The source of organic sulfur in these deposits has
been attributed to the participation of thiols (mercaptans),
sulfides, and disulfides (Taylor and Wallace, 1967).

164

Table 6.2 Representative Sulfur-containing Compounds in
         Crude Oil

Thiols (Mercaptans)   RSH

Sulfides   RSR'

Cyclic Sulfides

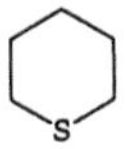

Disulfides   RSSR'

Thiophene

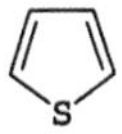

Benzothiophene

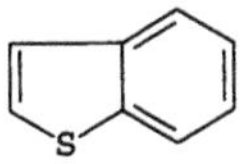

Dibenzothiophene

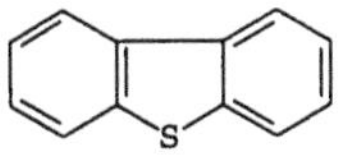

Napthobenzothiophene

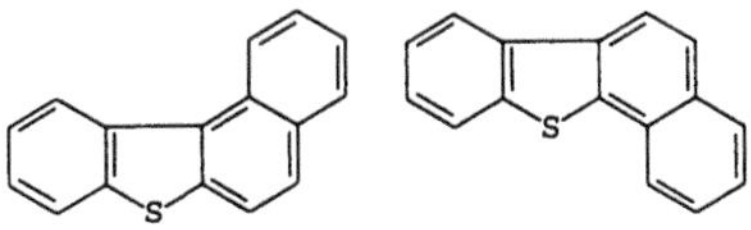

In jet fuels that have been deoxygenated, sulfides and
disulfides have been found to lead to increased solids formation
(Taylor, 1976). It has been shown that jet fuels low in sulfur
content are relatively stable and that fuels of high sulfur content
are fairly unstable. Deposits formed in jet fuel in the presence
of oxygen contain much greater percentages of sulfur, up to a 100-
fold increase, than that present in the fuel itself (Taylor and

165

Table 6.3. Ranges of Pricipal Trace Elements Found in Petroleum

Element	Range in petroleum (ppm)
Cu	0.2 - 12.0
Ca	1.0 - 2.5
Mg	1.0 - 2.5
Ba	0.001 - 0.1
Sr	0.001 - 0.1
Zn	0.5 - 1.0
Hg	0.03 - 0.1
Ce	0.001 - 0.6
B	0.001 - 0.1
Al	0.5 - 1.0
Ga	0.001 - 0.1
Ti	0.001 - 0.4
Zr	0.001 - 0.4
Si	0.1 - 5.0
Sn	0.1 - 0.3
Pb	0.001 - 0.2
V	5.0 - 1500
Fe	0.04 - 120
Co	0.001 - 12
Ni	3.0 - 120

Wallace, 1967). Historically, by contrast, it has been demonstrated that sulfur compounds present in lubricating oils act as antioxidants by a mechanism involving the decomposition of peroxides (Thompson et al., 1949; Walters et al., 1949). Thus, the presence of sulfur compounds in fuels is simultaneously desirable and deleterious. The rates of reactions in oxidation schemes are dependent on hydrocarbon structure, bond type, heteroatom concentration, heteroatom specification, temperature, and oxygen concentration (Turney, 1965; Hucknall, 1974). In oxidation schemes in fuels, the primary oxidant is always a hydroperoxide compound. Catalysts and free radical inhibitors can materially alter both the rate of oxidation and the reaction pathways (Mayo, 1972).

If sufficient oxygen is present, the hydroperoxide content will reach a high concentration. If the available oxygen is low but the temperature is raised, the hydroperoxide concentration will be limited by free radical decomposition. Under these conditions, fuel degradation can be associated with the formation/decomposition pathways of hydroperoxides and their reaction pathways with other heteroatom moieties and/or species such as olefins.

The detailed mechanism of hydroperoxide decomposition is complicated, since free radicals in a fuel matrix are sensitive to slight changes in their chemical environment. In these complex mixtures, it is difficult to identify specific reaction pathways. Co-oxidation and condensation reactions could offer possible explanations for the increased incorporation of heteroatoms observed in the deposits formed by fuels (Mushrush et al., 1994).

Chemically, in fuels of high nitrogen content (such as those derived from oil shale or coal) the nitrogen compounds, rather than the sulfur heteroatom content, are probably responsible for fuel degradation reactions (Cooney et al., 1984). Pyridine compounds, mostly long-alkyl-chain substituted, have been identified as a major nitrogen species present in fuels. Pyridines, alkyl or aryl substituted, rapidly develop intense colors after brief exposures to light and air.

On exposure to light and air at room temperature, pentadecyl pyridine rapidly develops into a deep green-black solution. Mass spectral analysis of the discolored pyridine solution showed the same purity as freshly distilled alkyl pyridine (Mushrush et al., 1987). Alkyl-substituted pyridines generally form black-colored solutions on short exposures to air. Oligomers of short-alkyl-chain-substituted pyrroles could also be involved in color body formation. A type of charge-transfer complex could be invoked to explain these color bodies.

Color bodies in and of themselves do not predict instability but could mask instability reactions, and fuel containing them would appear undesirable.

2.0 Functional Group Chemistry

Model compound studies have provided much of the information relating to the chemistry of instability/incompatibility. Such studies provide a means of identifying the compound types involved in color/sludge/sediment formation. The studies also allow deductions to be made about the potential interactions between the different molecular types. These studies then lead to determining the effects of storage conditions and, hopefully, shedding some light on the mechanisms of color/sludge/sediment formation. Coupled with the methods of testing for stability/incompatibility (Chapter 7) these studies will then offer some insights into the potential for instability/incompatibility in liquid fuels and other products.

However, valuable though the model compound studies can be, there is an additional factor that is not often taken into consideration. It is essential to recognize that in a mixture as chemically complex as many liquid fuels and other products can be, there is the potential for interference in the chemical reactions by other members of the mixture. It may be that the reactivity of the "interfering species" may not be recognized from reactivity studies of the individual constituents.

2.1 Oxygen

Practically all functional groups involving oxygen are present in hydrocarbon fuels. For the most part, aldehyde, ketone, alcohol, ester, and ether linkages have been shown to be innocuous.

In petroleum- and shale-derived fuels, phenols are present in relatively low concentrations and can function as an antioxidant and, thus, serve a useful role. In coal-derived fuel liquids and in some synfuels, phenols have been observed in high concentrations. In these instances, the phenols themselves can undergo oxidative coupling and participate in a major fashion in sediment formation.

2.1.1 Carboxylic Acids

Carboxylic acids have been found to be moderately detrimental in both fuels and model fuels. Carboxylic acids are implicated in incompatibility processes because they catalyze condensation reactions and thus rapidly increase both polarity and molecular weight.

However, the most deleterious oxygen functional group is the hydroperoxide. Peroxide species form by the reaction of dissolved oxygen. It is observed with time that the aldehydes, ketones, and carboxylic acids increase in concentration. These compounds are the secondary reaction products from hydroperoxides.

2.1.2 Hydroperoxide Reactions

This chapter deals primarily with hydroperoxides and their subsequent reactions. Thus, a cautionary note is in order; all peroxides should be considered unstable. Even with precautions, unexpected and unexplained explosions can occur at moderate temperatures.

It is probable, but not proven, that free radical reaction pathways can be altered with a rapid evolution of low molecular weight gases resulting in large pressure surges. Reactions that involve anhydrous peroxides with fuels or in the presence of metals or metal ions should be approached with extreme caution. It is prudent to minimize the use of anhydrous peroxides or hydroperoxides as reactants.

A hydroperoxide is the primary oxidant present in any hydrocarbon fuel. Increasing peroxide concentration is usually a predictor of the instability reactions to follow. The natural antioxidants present in fuels, primarily organic sulfur compounds keep peroxide concentration in check for short periods of storage time, but are eventually overwhelmed. The ability of these compounds to serve the role of antioxidants is probably what accounts for the induction period that is noted in instability reactions.

The stability of all fuels could be radically improved by removing oxygen or lowering its concentration to less than 3 ppm. However, this would engender transportation and storage problems that might be too great economically to overcome. Economical methods for lowering the dissolved oxygen content in storage is a fruitful area for further research. Nitrogen sparging, while easy in practice, would represent a large economic and environmental control burden, but would probably be the cheapest of the alternatives.

Alkyl hydroperoxides are liquids, with the lower members being soluble in both polar and nonpolar solvents. Low-molecular-weight hydroperoxides are explosive, but their thermal stability increases with increasing carbon number. Thus, the more stable peroxide compounds have a lower oxygen content. Alkyl hydroperoxides are stronger acids than the alcohols and, consequently, can be isolated through their Group IA salts. Homolytic, heterolytic, and oxidative cleavage of the peroxide bond can occur.

In addition, peroxide formation is noted in many hydroperoxide reactions. In a basic medium, hydroperoxides give carbonyl compounds. Under acidic conditions, a few hydroperoxides have been observed to undergo rearrangement in addition to the traditionally accepted reaction pathways.

The autoxidation of hydrocarbons below 200°C (392°F) leads to extensive hydroperoxide formation. Hydroperoxides themselves are the main reactant at higher-temperature regimes for subsequent autoxidation reactions. The O-O bond is the weakest bond in an alkyl or aryl hydroperoxide. The bond energy of the RO-OH bond is about 175 kJ/mole, with thermal homolysis giving alkoxy and hydroxy radicals (Benson and Shaw, 1970). At temperatures of about 100°C (212°F) alkyl hydroperoxides (but not allylic or aryl) are stable in hydrocarbon solvents, and it is the solvent itself that undergoes oxidation (Moshher and Durham, 1960).

When used as solvents, tetralin and cumene, for example, are oxidized to tetralin and cumene hydroperoxides. The thermal stability of tertiary alkyl hydroperoxides does not depend to any significant extent on the alkyl group present; thus t-butyl hydroperoxide and cumene hydroperoxide have similar chemical and thermodynamic stabilities (Hiatt and Strachan, 1963).

The stability of hydroperoxides is dependent on both the chemical environment and its concentration in a system. Stability based on chemical environment is more apparent than stability based on concentration. At low concentration, $<10^3$M of hydroperoxide, the reaction is first order in hydroperoxide, but at higher concentrations, $>10^1$M, it is second order with allylic hydroperoxides because of the following reaction sequence:

170

[1]
$$H-O-O-R$$
$$R-O-O-H$$
$\longrightarrow$ RO + H_2O + ROO

The peroxide hydrogen bond of an alkyl hydroperoxide is weak and the reaction with other radicals in the system is exothermic. In high concentrations, hydroperoxides are susceptible to induced decomposition by abstraction of the hydroperoxide hydrogen. However, the alkyl hydrogens of tertiary groups are practically inert because they are primary hydrogens that are not activated by the presence of any other groups.

Primary and secondary alkyl hydroperoxides react quite differently. Secondary, unlike tertiary hydroperoxides do not undergo an induced decomposition because secondary peroxy radicals undergo the following nonradical self-reaction:

[2]   2R2CHOO $\longrightarrow$ $R_2C=O$ + O_2 + R_2CHOH

Induced decomposition of hydroperoxides is catalyzed by catalytic amounts of metal ions. Trace amounts of metal ions are always present in fuel liquids. The reactions of hydroperoxides with metal ions have been well studied, but the synergism in complex systems such as fuels has been avoided.

The reactions leading to the formation of hydroperoxides under mild conditions in fuels have been extensively reported (Coordinating Research Council, 1979). The systematic investigations of aliphatic hydrocarbon oxidation fuel mechanisms were undertaken in the 1930-1950 time period. Under mild conditions, the reaction is actually a chain process. If the propagation steps are greater than those of termination, then alkyl radicals will react with dissolved oxygen in the system reaction steps [3] - [12] (Figure 6.1).

Reactions [3] and [4] are important in starting the oxidation process but do not contribute significantly to the ROOH concentration. In an oxygen-starved system, oxygen <5 kPa, step [4] becomes rate controlling.

The rate constant for reaction [4] has been reported to be $k = 10^7 - 10^9 M^1 sec^1$ (Denisov, 1974). Consequently, if the oxygen

171

concentration is at least 10^4M, then the peroxy radical ROO is present in much higher concentration than the alkyl radical, R . In the reaction mechanism (Figure 6.1) the peroxy radical is the least reactive of all the radicals generated. Thus, a significant concentration of the alkyl peroxy radical is available for further reaction. In a pure fuel, with no peroxide species present, the chain process will not begin until a trace of the hydroperoxide radical forms.

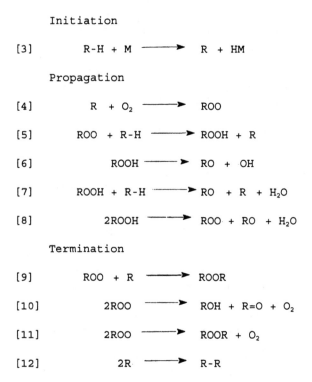

Initiation

[3]   R-H + M ──────► R + HM

Propagation

[4]   R + O_2 ──────► ROO

[5]   ROO + R-H ──────► ROOH + R

[6]   ROOH ──────► RO + OH

[7]   ROOH + R-H ──────► RO + R + H_2O

[8]   2ROOH ──────► ROO + RO + H_2O

Termination

[9]   ROO + R ──────► ROOR

[10]   2ROO ──────► ROH + R=O + O_2

[11]   2ROO ──────► ROOR + O_2

[12]   2R ──────► R-R

Figure 6.1   General Hydroperoxide Reaction Sequence

2.2 Sulfur

Several excellent reviews of the literature are available on both the types of sulfur species present in fossil fuels and the organic chemistry of sulfur compounds (Reid, 1958; Danehy, 1968; Kharasch, 1961; Oae, 1977; Hartough, 1964; Field, 1977).

[13] $RSCH_2CH_3$ + $t-C_4H_9O$

[13a] $RSCHCH_2$ (I) + $RSCH_2CH_2$ (II)

[13b] $-CH_2=CH_2-$ (III)

[13c]   RS· (IV)

Products include

[13d] addition products (I) + (I)

  ──────► $(RSCHCH_3)_2$ 32% yield

[13e] addition products (I) + (II)

  ──────► $RSCH_2CH(CH_3)SR$ 3.6% yield

[13f] addition to multiple bonds (III) + (IV)

  ──────► $(RSHCH_2)_2$ 2.2% yield

[13g] addition products (I) + (IV)

  ──────► $RSCH(CH_3)SR$ 2.9% yield (Migita et al., 1973)

Figure 6.2   Organo-sulfide Reaction Pathways

The reactions of the various sulfur functional groups under oxidizing conditions have been studied exhaustively. However, the influence of the intermediate oxidation products such as sulfoxides, thiolsulfonates, and the final product, a sulfonic acid, has not been studied in detail.

It is probable that these intermediates and the final oxidized product hold the key to both sediment formation and the instability of middle distillate fuels.

Many studies on sulfur compound oxidants, H_2O_2, O_2, ROOH, I_2, $KMnO_4$, SO_2, Fe^{+3}, H_3NO, Cr^{+3}, H_2SO_4, peracids, etc., have been reported, and the fate of sulfur-centered radicals under oxidizing

Table 6.4 Mole Percent Conversion of Hexyl Sulfide at
120°C with t-Butyl Hydroperoxide in Benzene Solvent

Sulfide Products	Mole % Yield With Time In Minutes				
	15	30	60	120	180
hexyl sulfoxide	74.8	85.9	81.6	81.9	80.7
hexyl sulfone	1.2	1.4	3.0	3.4	4.0
hexyl disulfide	0.3	0.5	0.5	0.5	0.6
hexanal	0.2	0.2	0.3	0.2	---
hexane	0.1	0.1	0.3	0.4	0.3
hexene	0.1	0.1	0.1	0.3	0.2
hexyl thiolsulfinate	0.1	0.1	0.1	0.1	0.1

conditions is known (Figure 6.2). Sulfides and thiols under oxidizing conditions produce the thiyl (RS·) radical. The fate of thiyl radicals is disulfides. Disulfides under mild or forced oxidizing conditions produce ultimately a sulfonic acid. Other acidic species that might be proposed include sulfenic and sulfinic acids.

Sulfenic acids are generally unstable, while sulfinic acids are usually prepared through the ester by moderately strong oxidants. Under mild oxidizing conditions, neither of these species is commonly observed.

2.2.1 Sulfides

Sulfides were first considered inhibitors to deposit formation in gasolines. Deposits formed in jet fuel in the presence of oxygen contained a greater percentage of sulfur than that present in the fuel itself. A similar result was observed for deoxygenated fuel doped with sulfides. Other reported results for sulfides are difficult to interpret.

It has been reported that in Jet A fuel that has been doped with butyl sulfide, the amount of deposit formation differs when the same fuel is doped with pentyl sulfide. Bond energies for these two sulfides are practically identical. Unless impurities are present in the reagents, these two sulfides should give identical results. Experience shows that most sulfides contain small amounts of thiols. It might be suggested that the observed differences are due to the analytical method employed.

A wide range of chemical interactions can occur with sulfides, including but not confined to abstraction, fragmentation, combination, dimerization, addition to unsaturated bonds, displacement, and rearrangement. From the perspective of fuels, abstraction, combination, and addition to unsaturated bonds are the most probable reaction pathways, with radical annihilation by dimerization the lesser pathway. This is the problem with comparing model studies to systems employing fuels. Model studies have a limited slate of radicals and alternate pathways available for reactions [13] - [13(g)] (Figure 6.2).

The sulfur compound product slate for a model study involving the reaction of t-butyl hydroperoxide with hexyl sulfide in de-aerated benzene solvent is varied, although the major product observed was hexyl sulfoxide (Table 6.4). Other products included hexyl sulfone and hexyl disulfide. Minor products included hexyl thiolsulfinate, hexanal, hexene, and hexane.

The hexyl sulfoxide could result from several mechanisms, the most likely being the reaction of the t-butyl hydroperoxide itself with the hexyl sulfide, followed by a rapid proton transfer and O-O bond rupture (Rahman and Williams, 1970).

Expansion of the sulfur valence shell is probable in the processes involved in this step. Another mechanism could involve the attack of an oxygen-centered radical, i.e., peroxy on sulfur followed by a beta-scission (Bridgewater and Sexton, 1978).

In the presence of the very reactive peracids or peresters, sulfoxides and small amounts of the sulfones are formed (Modena and Todesco, 1962). Only in the case of powerful peracids, i.e.,

175

trifluoroperoxyacetic acid, do sulfones form directly (Liotta and Hoff, 1980). The alkyl sulfone product then forms by a similar mechanism. The nucleophile attacks the sulfoxide as an S-electrophile.

Peracids and esters on subsequent reaction ultimately form sulfonic acids from both aryl and alkyl sulfides.

An interesting case involving a substituent effect is that of phenyl alkyl sulfides. Aryl sulfides react and degrade on the basis of substituent effect, i.e., a sulfonyl group protects the phenyl moiety, while an alkyl group is oxidized to a carboxylic acid (Curci et al., 1966).

Sulfoxides, once formed, are quite stable under mild oxidizing conditions. The major hexyl sulfoxide oxidation product, hexyl sulfone, varied from an initial 1.2% at 15 min to 4.0% at 180 min of reaction. This slight increase is comparison with the yield of hexyl sulfoxide at 180 min illustrated the resistance of an alkyl sulfoxide to oxidation by a hydroperoxide. These partially oxidized sulfur species, however, are not as stable as the starting hexyl sulfide. Oxidation of a sulfoxide to a sulfone is believed to proceed by a mechanism similar to that for sulfoxide formation.

The formation of alkyl sulfones is a facile reaction only in the presence of a strong oxidant or when the reaction is catalyzed by transition metal ions (Henbest and Khan, 1968).

2.2.2 Thiols and Disulfides

The thiyl radical RS· is the most reactive of the sulfur-centered radicals that can form in an oxidizing medium. It can be generated from most sulfur species, but thiols serve as a convenient source. This sulfur-centered radical behaves quite differently from the corresponding oxygen-centered radical, the alkoxy radical. Alkoxy radicals produce both the cleavage product, a ketone, as well as the hydrogen abstraction product, an alcohol. The chemistry of alkoxy and thiyl radicals is, however, quite different, as illustrated by reactions [14] and [15]:

[14] $RS· + R-H \longrightarrow RS-H + R·$

[15] $(R)_3RS· \longrightarrow (R)_2R=S + R·$

For the thiyl radical, neither the thiol, to any significant extent, nor the thione have been observed to form by these reaction pathways. For reaction [14], the reason is thermodynamic in nature. The reverse reaction is, in theory, favored by several kcal/mole. Only in cases involving very stable radicals is reaction [14] observed. In the case of reaction [15], thiones have never been observed from the simple thiyl radical cleavage (Van Swet and Kooyman, 1968).

The mechanism of thiol oxidation has been the subject of discussion for many years (Reid, 1958). In the presence of oxygen, alkyl thiols are unreactive, with t-alkyl thiols being practically inert. Heavy metal ions, however, catalyze the oxidation of thiols. This is an area that is ripe for study: the interaction of metal ion catalysis of the hydroperoxide-initiated oxidation of sulfur species. In the presence of oxygen, alkyl thiols react through an intermediate hydrogen peroxide (Ohno, 1971):

[16] $RSH + O_2 \longrightarrow RSSR + H_2O_2$

Results from the t-butyl hydroperoxide oxidation of alkyl thiols support the thesis that the t-butoxy or peroxy radical generated from t-butyl hydroperoxide abstracts the thiol hydrogen:

[17] $RS\text{-}H + t\text{-}C_4H_9O\cdot \longrightarrow RS\cdot + t\text{-}C_4H_9OH$

Rapid dimerization, equation [18], then follows generating the major product, an alkyl disulfide:

[18] $2RS\cdot \longrightarrow R\text{-}S\text{-}S\text{-}R$

Historically, in the presence of oxygen, the thiyl radical adds to olefins to give hydroxy sulfoxides as the major product and the sulfide as a minor product:

[19] $RC\cdot HCH_2SR + O_2 \longrightarrow RCHCH_2SR$
$$O\text{-}O\cdot$$

In an acid-wash refining step, thiols can be oxidized to eventually yield very reactive sulfur species, such as disulfides and sulfur dioxide:

[21] $2RSH + H_2SO_4 \quad \cdots\cdots\longrightarrow RSSR + SO_2 + 2H_2O$

Sulfur dioxide then reacts with hydroperoxides, reaction [22] in the system to produce a very reactive perester of a sulfonic acid:

[22] $ROOH + SO_2 \quad \cdots\cdots\longrightarrow ROSO_2OH$

Reactions involving simultaneous addition and oxidation reactions are quite common both in model systems and in fuels. A reaction sequence involving thiols with active olefins, such as styrene and indene, in the presence of hydroperoxides or molecular oxygen, demonstrates these two co-oxidation processes.
Whether the disulfide is a result of the oxidation reactions of thiols or was originally present, the ultimate oxidation product is a sulfonic acid:

[23] $RSSR + [oxidant] \quad \cdots\cdots\longrightarrow$ [23a] + [23b]

[23a] $\cdots\cdots\longrightarrow RSOH \qquad \cdot\longrightarrow RSOOH \qquad \longrightarrow RSO_3H$

            sulfenic acid $\longrightarrow$ sulfinic acid

[23b] $\longrightarrow RS(O)SR \quad \longrightarrow RS(O)_2SR \quad \longrightarrow RS(O)_2S(O)R \quad \longrightarrow RS(O)_2S(O)_2R$

Reaction pathway [23] illustrates the reaction scheme. Reaction [23a] illustrates the probable path to a sulfenic and sulfinic acid. In reaction [23b], a thiolsulfonate is stable if the R groups are long alkyl chains, but unstable if they are short-chain or aryl groups. With short chains, disproportionation to disulfides and thiolsulfonates is observed. Thiolsulfonates are quite stable in contrast to sulfinyl sulfones. Disulfones are stable under ordinary conditions, but can be cleaved by nucleophiles. Under rigorous oxidizing conditions, a sulfonic acid is the expected product, but the situation is less clear under mild conditions.
The generation of acidic species has important implications for the general area of fuel instability. Literature reports implicating sulfonic acids in a catalytic role are increasing

(Hazlett et al., 1991). Sulfenic and sulfinic acids are not usually formed by mild oxidation with hydroperoxides; thiolsulfinates, thiolsulfonates, and disulfones are observed under these conditions. Thus, it is important to look at the sulfonic acid class of compounds present in both the crude and the finished fuel to understand the mechanism of fuel instability processes.

Model fuel studies can give misleading results with systems that are propagated by the generation of free radicals. Sulfur-centered radicals, under mild reaction conditions, can be quite high in concentration in both real fuels and model systems. Studies with high radical concentrations are usually conducted in solvents that are not susceptible to radical attack, thus leading to an abnormally high yield of sulfur radicals undergoing self-reaction. On the other hand, sulfur-centered radicals generated in a fuel matrix would be expected to be highly reactive, with no appreciable steady state concentration under ambient conditions.

The rapid disappearance of sulfur-centered radicals can be attibuted to the dimerization reaction, which at ambient conditions is thermodynamically favored ($\Delta H = -73$ kcal/mole). This process is generally assumed to proceed at the diffusion-controlled rate constant. Reaction [18] illustrates the process. More probable would be combination reactions with olefins in the system, hydrogen abstraction reactions, and radical rearrangement.

The reactions of any sulfur compound can then produce a thiyl radical that by self-annihilation leads first to a disulfide but eventually to a sulfonic acid. Thus, it is important to examine each functional class of sulfur compounds for sulfonic acid potential.

2.2.3 Thiophenes

Substituted and condensed thiophenes are usually the most abundant heteroaromatic sulfur species in refined fossil fuels (Kong et al.,1984). There are several rapid and simple methods for the determination of both condensed thiophenes and other types of aromatic sulfur compounds based on ligand-exchange chromatography coupled with gas chromatography/mass spectroscopy. Condensed thiophenes, but not dibenzothiophenes, have been found to be deleterious in terms of fuel stability. The thermal and reactive stabilities of the thiophenes approach those of aromatic hydrocarbons, due to their resonance stability energies.

The similarity to aromatic hydrocarbons also makes the direct gas chromatography/mass spectroscopy identification of these sulfur species difficult in a fuel matrix. Thiophene compounds reported in the literature with reference to fuel instability have included thiophenes, tetrahydrothiophenes, alkyl-substituted thiophenes, and condensed thiophenes such as benzothiophenes and dibenzothiophenes. Oxidation reactions with peroxide species and the thiophenes, alkyl substituted thiophenes, benzothiophenes, and dibenzothiophenes have been reported.

A thiophene ring system can survive moderate oxidizing conditions, but side reactions involving condensation must be anticipated (Figure 6.3). However, if the ring is attacked, it generally breaks down to oxalic or other substituted carboxylic acids, with the sulfur being converted to sulfuric acid. The literature contains little information on the mild oxidation of

Table 6.5: Mole % conversion for tetrahydrothiophene and t-Butyl Hydroperoxide.

Products Derived from tBHP	Reaction Time (Min)				
	15	30	60	120	180
t-butanol	34.1	42.6	55.2	62.2	68.6
acetone	1.3	2.2	2.2	2.3	2.3
methane	0.8	1.1	1.2	1.2	1.4
isobutylene	3.4	3.0	3.0	2.9	2.6
di-t-butylperoxide	0.3	0.5	0.5	0.7	0.7
Tetrahydrothiophene					
sulfoxide	13.8	22.8	23.8	25.7	26.8
sulfone	----	0.1	0.2	0.2	0.2
Unreacted					
tBHP	52.3	44.2	35.7	20.1	10.3
tetrahydrothiophene	85.2	76.6	74.5	73.0	71.4
Trace Products	0.2	1.1	1.7	2.0	3.6

thiophenes, especially in a radical environment. In the presence of low concentration of hydroperoxides, no evidence of ring opening was observed (Mushrush et al., 1987).

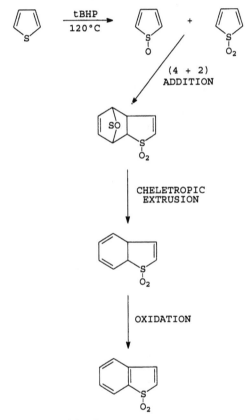

Figure 6.3: Thiopene oxidation.

Even though it is not possible to detail the actual mechanism for thiophene oxidation, speculation based on product distribution is possible. It is well known that both the sulfoxide and sulfone of thiophene itself are only stable when the thiophene is highly substituted. The oxidation of thiophene and its methylated analogues does not yield simple, easy-to-identify products.

Detailing the reaction mechanism for thiophenes is further complicated, in that the solid products are not thermally stable. Products from both thiophene and 2,5-dimethylthiophene included traces of the sulfoxide, the sulfone, and water soluble Diels-Alder condensation products, 4:7:8:9-tetrahydro- 4:7-sulfinyl thionaphthene 1:1-dioxide (I), and thionaphthene 1:1-dioxide (II),

Table 6.6: Nitrogen compounds analysis up to 550 amu for typical shale and petroleum derived jet fuels

Nitrogen Compound Class	% Nitrogen Compounds		
	Shale		Petroleum
	Isomers	%	%
Carbazoles	98	1.3	0.5
Indoles	87	3.7	9.1
Pyridines/anilines	388	73.5	64.3
Pyrroles	42	0.6	1.1
Quinolines	66	4.1	10.4
Tetrahydroquinolines	248	9.9	3.2

along with the 2,5-dimethyl analogues. The 2,5-Dimethyl thiophene was reported to be more reactive than thiophene when measured by thiophene compound disappearance.

Results for the oxidation of tetrahydrothiophene by t-butyl hydroperoxide have been reported (Table 6.5) (Mushrush, 1991). This reaction gave the sulfoxide as the major product and a trace amount of sulfone. From t-butyl hydroperoxide, t-butanol was the major product, with small amounts of acetone, isobutylene, methane, and di-t-butyl peroxide products included: tetrahydrothiophene, sulfone, methane, and isobutylene. Thus, this cyclic sulfide gave a product slate similar to an aliphatic sulfide (Table 6.4), although the yields were much lower.

Substituted thiophenes, such as benzothiophene, are the least reactive of the thiophenes under mild oxidizing conditions that would be encountered under ambient storage conditions. Under

182

oxidizing conditions, when benzothiophene reacts with hydroperoxides, at the end of a 3-hr. reaction, more than 97% of the
benzothiophene remained unreacted.   In the presence of a
hydroperoxide, benzothiophene yields minor amounts of the sulfone
product and a trace of the sulfoxide product.

2.3 Nitrogen

Polar nitrogen compounds have two deleterious affects on crude
oil and on fuels made from these crude oils.   They poison oil
refining catalysts and affect the quality of finished fuels, since
they can be implicated in product instability.   Present knowledge
has suggested that for some fuels, nitrogen heterocycles may play
a causative role in the formation of insoluble sediments and gums
under conditions of ambient and accelerated storage (Cooney et al.,

Table 6.7: Relationship of sediment formation with organic
nitrogen compound structure

Group 1 (> 100mg/100mL)	Group 2 (10-100mg/100mL)
dodecahydrocarbazole	3-methylindole
2,5-dimethylpyrrole	2,5-dipropylpyrrole
2,4-dimethylpyrrole	pentamethylpyrrole
1,2,5-trimethyl pyrrole	2,3-dimethylindoline

Group 3 (3-10mg/100mL)	Group 4 (< 3mg/100mL)
2-methylpyrrole	2,6-dimethylpyridine
3-methylisoquinoline	3,5-dimethylpyrazole
1-methylpyrrolidine	2-methylquinoline
2,6-dimethylquinoline	2-methylpyrazine
2-benzylpyridine	2-methylpyridine
2-methylpiperdine	2-methylindole
pyrrolidine	3-methylpyridine
2,6-dimethylpyridine	4-methylpyridine
4-isobutylquinoline	4-methylquinoline
pyrrole-2-carboxaldehyde	1,2,3,4-tetrahydro-quinoline

1984). The nitrogen compounds that occur in finished fuels can be conveniently divided into nonbasic and basic fractions. The nonbasic nitrogen compounds include carbazoles, indoles, pyrroles, and quinolines. The basic fraction would include the pyridines and amines.

In particular, nitrogen-containing aromatics (such as pyrroles, indoles, and carbazoles) appear to be very harmful. Nitrogen heterocyclic compounds, which commonly occur in fuels, include alkylated pyrroles in the gasoline fraction along with alkylated pyridines, quinolines, tetrahydroquinolines, indoles, pyrroles, and carbazoles in middle distillate fuels.

Jet fuels contain a wide range of nitrogen compounds (Table 6.6). This particular shale-derived fuel was subjected to severe hydrotreatment, while the petroleum fuel was a composite of 70% straight-run and 30% hydrocracked streams. Both of these fuels were marginally stable as defined by the formation of solid products (Mushrush and Hazlett, 1985).

Fuel instability is reflected by the concentration of the polar nitrogen species present. However, it is not only the basic (pyridine-type) nitrogen that is deleterious to operations, but also the carbazoles, indoles, and pyrroles appear to be much more deleterious than other classes of polar nitrogen species. Practically all reports of nitrogen class distribution in hydrocarbon liquids show that shale- and coal-derived fuels are higher in indoles, pyridines, and carbazoles, while petroleum fuels tend to be lower in concentration in these functional groups. The gasoline distillation range appears to have a higher concentration of pyrrole-type nitrogen structures. The various nitrogen functionalities can be correlated with sediment formation in middle distillate fuels (Table 6.7).

Currently, it appears that relatively nonbasic nitrogen heterocyclics, particularly those that contain alkyl groups in certain isomeric positions, may be the most troublesome compounds. Thus, stereochemistry is quite important in determining the course of reactions in fuels.

Since both carboxylic and sulfonic acids are implicated in instability reactions, the structure of the nitrogen compound would be expected to play a major role. Increasingly, we are led to believe that interactive processes between the basic and the nonbasic nitrogen moieties are of major significance in interactive

184

processes.

2.3.1 Pyridines

Pyridine compounds are thermally quite stable (Joule and Smith, 1979). Since they are easily protonated, they might seem to be likely candidates to be implicated in fuel instability reactions. However, pyridine compounds have not been observed in mass spectral analysis of any of the sediments. When pyridine compounds are used in model systems (Table 6.7), they do not induce further instability reactions. Most pyridine compounds either fit into Group 4 or do not induce any sediment to form. Thus, they do not serve either as a reactant or a catalyst for further instability processes (Cooney et al., 1984).

The actual compounds that are found in middle distillate fuels usually have alkyl substituents in multiring positions. The carbon number of the alkyl chain is dependent on the fuel distillation range.

Pyridine compounds are easily oxidized by strong oxidants to the corresponding pyridine N-oxide. Oxygen and hydroperoxides are not capable of oxidizing pyridine under ambient storage conditions,

Table 6.8 Characteristics of DMP Sediment

Analysis	Findings
IR	N-H
	C=O
	CH_2 + CH_3
C^{13} NMR	N-H
	C=O
	C-O ($\nearrow$O)
	CH_2 + CH_3
Elemental	$C_{6.3}H_{7.4}NC_{1.7}$

but at the considerably higher temperatures of nozzles or injectors, both of these species could oxidize pyridines.

Substitution position on the pyridine ring markedly affects reactivity. Alkyl or alkenyl isomers in the 2- or 4-position result in compounds that are much more acidic than for the 3-isomer. These 2- and 4-isomers undergo the usual condensation reactions. However, in the 3-position, the resultant pyridines are quite stable.

Pyridines containing an active olefin group, i.e., 2-vinyl pyridine, could be implicated in polymerization reactions. These compounds show many chemical similarities to styrene and other benzene homologs. The pyridine homologs are more reactive because of the labile N-H bond. Compounds like 2-vinyl pyridine would undergo reaction during the thermal refining and hydrotreating steps and not appear in the resultant fuels in any significant concentration. Alkyl-substituted pyridine compounds that have been subjected to hydrotreating yield alkyl-substituted piperdines. Piperdines show little tendency to react with weak oxidants.

2.3.2 Quinolines

Quinolines behave chemically in a similar fashion to pyridines. Model systems (Table 6.8) show that the quinolines fit in the least reactive category for sediment formation, a characteristic similar to pyridines.

On hydrotreating, quinolines are converted to dihydro-quinolines, tetrahydroquinolines, and decahydroquinolines. Dihydroquinolines can then undergo facile oxidation by air back to the respective quinoline.

In slightly acidic solutions, and with no oxygen present, dihydroquinolines will disproportionate to yield tetrahydro-quinoline and quinoline. The quinoline to tetrahydroquinoline ratio is a measure of the severity of the hydrotreatment. Mass spectral analysis of sediments formed during storage or thermal instability regimes does not show any significant incorporation of the quinoline nucleus.

Quinoline is resistant to most mild oxidizing agents. Oxygen and peroxide concentrations commonly found in fuels apparently have little or no effect on the quinoline ring system. The use of stronger oxidants results in the formation of N-oxides. They follow the chemistry of the pyridines N-oxides, with one major exception, namely the presence of a benzene ring allows for electrophilic substitution. Electrophilic monosubstitution occurs at the positions adjacent to the ring junction. In this case a strong similarity to naphthalene chemistry is noted.

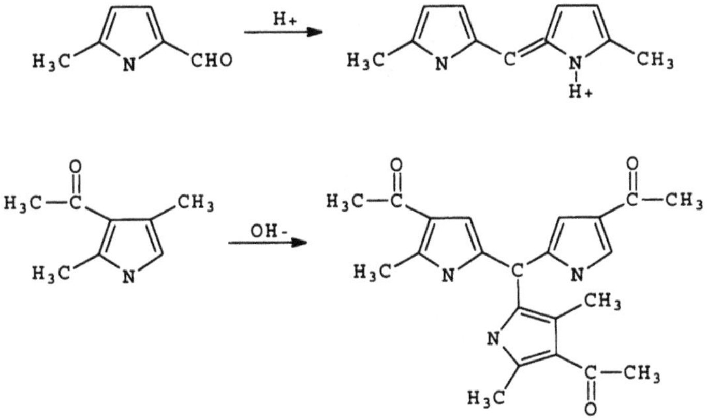

Figure 6.4   Acid Base Reactions of Substituted Pyrroles

Thus, temperature and reaction conditions both alter and govern the position of substitution. However, the position of the alkyl substitution does not seem to have any significant effect on reactivity with respect to instability.

2.3.3 Pyrroles
Pyrroles show some chemical similarities, but significantly more differences, with the pyridines and the other nitrogen functional groups. The major similarity is that of thermal stability. Differences include the ease of oxidation of pyrrole, and the facts that pyridine is basic and acid stable while pyrrole is practically neutral and readily polymerized, and that pyridine is resistant to electrophilic reagents while pyrrole readily reacts. The ease of polymerization is shown by the reaction of pyrrole dissolved in ether containing a few drops of acid: immediate polymerization results.

However, pyrroles that contain electron-withdrawing groups are either nonreactive or form simple salts with mineral acids. The

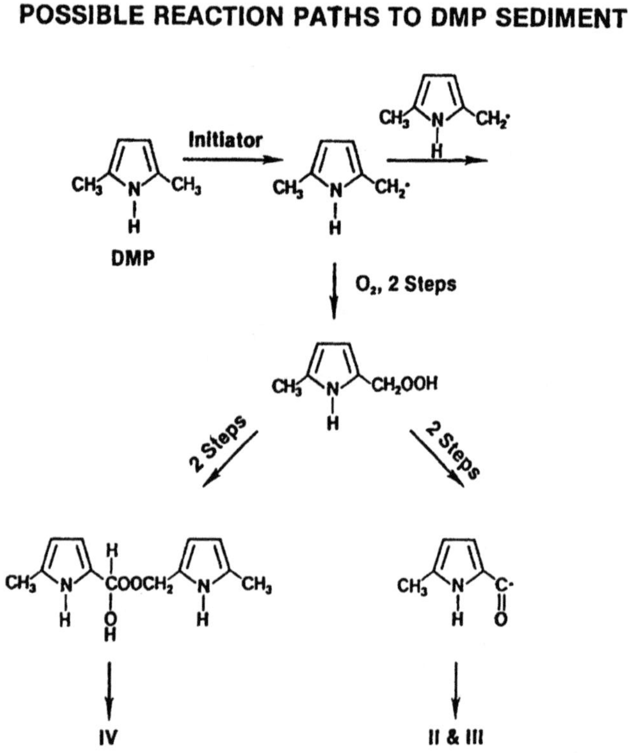

POSSIBLE REACTION PATHS TO DMP SEDIMENT

Figure 6.5: Possible reactions in the formation of sediment from dimethyl pyrrole.

oligomer product was referred to as "pyrrole black" in the early literature. The characteristics of the sediment formed from 2,5-dimethylpyrrole shows oxygen functions were that not present in the monomer (Table 6.8), leading to the inference that dimerization can occur through the methyl groups (Figures 6.4 and 6.5), although there is always the potential for dimerization (and hence, polymerization) to occur by other mechanisms. This infers then that the reaction of 2,5-dimethylpyrrole is the result of autoxidation processes and not a simple polymerization reaction.

The pyrrole oligomer decomposes on heating to pyrrole, indole and ammonia.

The ease of oxidation for pyrrole compounds is illustrated by the rapid darkening of freshly distilled pyrrole. Thus, the pyrroles that would be found in a finished fuel would be alkyl or aromatic substituted. This further suggests that feeds rich in pyrroles should probably not be subjected to an acid wash to remove basic nitrogen such as pyridines and amines. These basic moieties have been shown to be practically inert in finished fuels.

Most fuels show a relatively low concentration of pyrrole structures, most of those present being mono- or di-alkyl-substituted short chains, methyl through pentyl. Short-chain multisubstituted pyrroles readily form peroxides on exposure to air. Subsequently, these unstable peroxides would serve as a radical source for other oxidation reactions.

Dimethylpyrrole has been the subject of many model system instability studies. Dimethylpyrrole was probably chosen because the reaction itself was well defined in the literature. However, it bears little relationship to the actual process occurring in fuels. Sediments generated from 2,5-dimethylpyrrole are not sensitive to the type of fuel, reaction temperature, solvent or concentration. In short, these studies are a good presentation of the chemistry of 2,5-dimethylpyrrole and nothing more. When

Table 6.9: Structure and reactivity of indoles

Compound	Dimer and Oligomers Products	No Reaction	Extent of Reaction
Indole	X		97%
1-Methylindole	X		94%
2-Methylindole		X	
3-methylindole	X		92%
7-Methylindole	X		96%
3-Ethylindole	X		83%
3-n-Propylindole	X		62%
3-n-Butylindole	X		5%
3-isoPropylindole		X	
3-tertButylindole		X	

employed in instability studies, 2,5-dimethylpyrrole generated more sediment than any other nitrogen functional group.

Pyrroles readily react with aldehydes and ketones by a 1-4-addition reaction to give high-molecular-weight polymers. Aldehydes and ketones are the secondary reaction products from hydroperoxides. Thus, their concentration can build to relatively high levels during ambient storage. However, pyrroles are present in low concentrations so this reaction does not explain the preponderance of sediment formed.

2.3.4 Indoles and Carbazoles

Since carbazole can be considered a substituted indole, the two compound classes will be considered as similar in sediment generation (Table 6.7).

Indoles and pyrroles show similarities rather than differences. Similarities are that they are both slightly basic, are both resistant to substitution; and have similar reactivity (the "C_3" of indole and the "C_2" of pyrrole are the same order of reactivity). Both react readily with electrophilic reagents, and they both form dimers and trimers.

Table 6.10: Reactivity of 2-methylindole with other indoles under mildly acidic Conditions.

Indole Compound	Formation of Mixed Oligomers	No Oligomers
Indole	X	
1-Methylindole	X	
2-Methylindole		X
3-methylindole	X	
7-Methylindole	X	
3-Ethylindole	X	
3-n-Propylindole	X	
3-n-Butylindole	X	
3-isoPropylindole		X
3-tertButylindole		X

Indole forms dimer salts in ether solvent that is slightly acidic. With aqueous acids, a mixture of dimers and trimers form. Increasing polymerization is then the result of autoxidation reactions. Both indoles and carbazoles will readily form hydroperoxides. Autoxidation reactions readily occur with 2,3-dialkyl and 3-alkyl substituted indoles.

The 2-methylindole isomer is very stable, and the very slight degree of reactivity (less than 1%) is still initiated at the C-3 position. Dimers of 2-methylindole cannot be synthesized directly, but must be synthesized by a reduction of the autoxidation product and once formed, they readily dissociate into the monomers.

CHEMICAL PROCESSES	PHYSICAL PROCESSES
Fuel Gasoline Jet Diesel	**Coalescence** insoluble products 1000 - 5000 Angstroms
↓	↓
Soluble Macromolecular Intermediate initial oxidation intermediates fuel soluble increasing polarity heteroatom incorporation	**Surface Effects** Collect on walls and/or fuel lines
↓	↓
Insoluble Products fuel insoluble high heteroatom content Molecular weight >500	**Temperature Effects** extensive oxidation at high temperatures leads to "coking"

Figure 6.6 Chemical and physical processes that lead to insoluble products

Saturated carbazoles readily form a hydroperoxide. For example, both tetrahydrocarbazole and dodecahydrocarbazole form a similar hydroperoxide. Both the carbazole and indole hydroperoxides have a transitory existence and, subsequently, react further by a free radical process. These same compounds can form transannular peroxides on exposure to air.

Some of the indole hydroperoxide reaction products can be implicated in color body formation. The 3-hydroxyindole is a reaction product, derived from either 3-indole hydroperoxide or from the oxidation of 3-methylindole with oxygen, which shows interesting chemistry. In slightly acidic media, a free radical reaction occurs at room temperature in air that results eventually in formation of the dye, indigo.

Alkyl-substituted indoles show a wide range of reactivities. In slightly acid solutions, dimers and trimers of certain alkyl derivatives are formed. The 2-alkyl substituted indoles do not condense to form dimers and trimers, the 3-alkyl-substituted indoles readily form dimers and trimers, and alkyl substitution does affect the reactivity and product slate (Tables 6.9 and 6.10).

3.0 Application to Fuels and Other Products

Instability can be thought of as an interactive process involving four main functional groups (Table 6.11) in combination with various physical processes (Figure 6.6). The difficulty arises in isolating the importance of each functional group to the overall process. It is truly amazing that less than 1 ppm of the fuel itself is involved in the process that has the potential to be detrimental to a whole storage tank.

It is a reasonable contention that incompatibility of fuels can be explained by these four functional groups. The chemical reaction pathway would consist of the following steps: hydroperoxide oxidization of organic sulfur compounds to sulfonic acids, which then subsequently catalyze condensation reactions between the other functional groups present, leading both to incorporation of heteroatoms and a simultaneous increase in polarity and molecular weight and, thus, precipitation from the fuel (Figures 6.7 and 6.8). The exact chemical composition of the sediment will depend on the chemical composition of the particular fuel, and hence, each fuel will display a unique sediment.

However, certain compound classes are observed to be common to

Table 6.11: Functional groups involved in incompatibility
reactions.

R-C=C- +	R-N- +	R-S- +	R-O-O-
Benzenes	Indoles	Sulfides	Hydroperoxides
Naphthalenes	Carbazoles	Thiols	Oxygen
Indenes	Pyridines	Disulfides	
Polycyclics	Pyrroles	Thiophenes	
	Quinolines		

all sediments. Thus, the presence of these heteroatomic compound
classes can be used to predict incompatibility. All of these
processes depend on dissolved oxygen and/or hydroperoxides to
initiate the reaction. Removal of the oxygen and/or
hydroperoxides, usually by clay filtration, will improve fuel
compatibility dramatically.

These functional groups are involved in both chemical and
physical processes that lead to deleterious solids.

3.1 Reaction Mechanisms

In the sedimentation process, a major unknown is the identity
of the soluble precursor to these sediments. These precursors are
in the early stages of research and have been termed **soluble
macromolecular oxidatively reactive species** (Hardy and Wechter,
1990). If this intermediate could be chemically elucidated, the
sedimentation process observed in practically all fuels,
incompatibility would be better understood.

A matrix involving time, temperature, heteroatoms,
hydroperoxides, and other reactive species present in fuels in the
context of accelerated storage is complicated and not well
understood. Accelerated fuel stability tests are important to both
producers and users of fuels, so an understanding of these
interactions is important. Model dopant studies provide a method
for isolating some of these variables. Model studies to define
chemical incompatibility work well with gasoline and jet fuels, but
are somewhat less reliable with chemically more complicated diesel
fuels. With relatively simple fuel, model studies set the

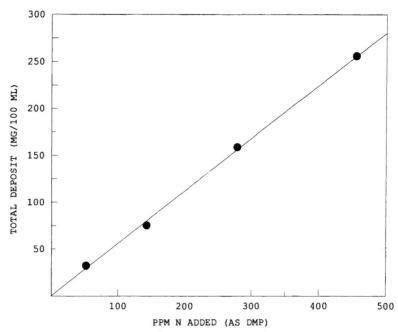

Figure 6.7: Total deposit in (m/100mL) with various DMP
Concentrations

parameters for incompatibility.

Thus, tests performed to mimic long-term storage give results
that are definitive as long as the temperature employed is low
enough not to initiate homolytic hydroperoxide reactions. The
majority of accelerated storage tests quantify sediment weight in
order to estimate fuel instability. Test temperatures have varied
from ambient to 150°C (300°F). Higher temperatures enable storage
tests to be completed in the minimum time, a producer advantage.

The significance of the accompanying uncertainty of the
observed sediment-producing processes in relation to the real
ambient storage conditions may be dominant, a user disadvantage.
Much of the early work has concentrated on the use of nitrogen
compounds, both basic and nonbasic, as dopants. It was
demonstrated that the solids formed from these dopants were similar

to the solids obtained from shale liquids themselves. The promotion
of sediment formation by these dopants has been reported to be a
facile process.

It has been proposed (Oswald and Noel, 1961) that the
sediments formed in middle distillate fuels were the result of
co-oxidation reactions between thiols, olefins, and pyrrole-type
nitrogen. In addition, it has also been proposed that the sulfonic
acids arose from the oxidation of aromatic thiols (Offenhauer, et
al., 1956).

Model oxidation studies have shown that the aromatic thiols

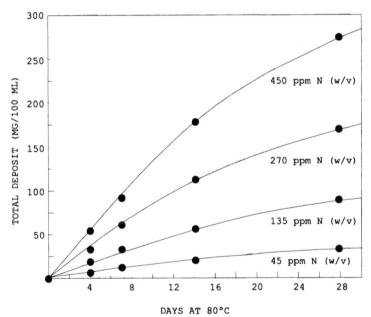

DAYS AT 80°C

Figure 6.8   Total deposit (mg/100mL) with various
concentrations of DMP for the 80° matrix

produced small amounts of sulfonic acid species under mild
oxidizing conditions. This chapter will deal with the interactive
chemistry that culminates in the gasoline, jet, and diesel fuel

instability processes. The feed stream, of course, will determine the quantity and type of heteroatoms present and, thus, stability of the finished fuel. The reaction mechanisms (above) are borne out by the results of interactive model studies with real fuels.

The reactions that cause instability in gasoline are much simpler than those in diesel and jet fuels. Diesel is much more complex chemically and, consequently, difficult to study. One must be careful in extending model studies to a real fuel environment. However, model systems present the only practical way to delineate the scope and mechanism of a particular reaction. Organic nitrogen compounds can be employed as dopants; for example, the results for 2-methylpyridine, 2,6-dimethylquinoline, dodecahydrocarbazole, 3-methylindole, and 2,5-dimethylpyrrole, added to a shale oil derived middle distillate that has the properties of a typical diesel fuel, show significant effects (Table 6.12).

The experimental results are practically the same when the same fuel contains an antioxidant. This is not surprising, since most additive packages used for middle distillate fuels were developed originally for gasoline. Their overall effect is somewhat in question with fuels that are much more chemically complex.

Table 6.12: Total insolubles for a shale-derived fuel with added codopants for a temperature/time matrix of 80°C/7 days.

Total Insolubles in mg/100mL Fuel					
Concentration of Dopant (ppm N)	MP	DMO	DHC	MI	DMP
0	0	0	0.2	0.1	0.1
45	0	0.2	46.6	1.7	7.9
135	0	0.2	75.3	1.5	25.4
270	0	0.1	158.7	4.0	58.1
450	0.1	0.1	322.9	12.3	97.3

3.2 Role of Heteroatoms

Heteroatoms are believed to play a major role in the formation of sediments, although they are not the only precursors to incompatible sediments. Olefins, aromatics, and other hydrocarbon species are active in sediment formation. Nevertheless, the removal of heteroatoms is recognized as a necessary step if compatibility is to be achieved. The major process for heteroatom removal is hydrotreating.

Hydrotreating a cracked distillate involves a formidable combination of reactions that occur simultaneously: hydrodesulfurization (HDS), hydrodenitrogenation (HDN), hydrodeoxygenation (HDO), hydrogenation of aromatics, and thermal cracking.

In addition the fate of basic and nonbasic compounds must be distinguished because they have different effects on downstream processes, for example, pyrrolic or nonbasic nitrogen has been implicated in sludge formation and light instability of gas oils (Chmielowiec et al., 1987). Temperature, pressure, residence time, and catalyst also vary from study to study. Given this number of variables, contradictions are bound to occur.

For example, carbazoles were more resistant than quinolines to hydrogenation (Dorbon et al., 1984), while at higher nitrogen content and temperature both basic and nonbasic nitrogen were persistent (Furimsky et al., 1978).

The reaction rates for HDN and HDS were both strongly dependent on the molecular weight and chemical characteristics of the feed oil (Trytten et al., 1990). At reaction temperatures of 400-425°C (750-800°F), nitrogen species in low-boiling fractions were much more reactive, and the basic pyrrolic compounds in the high-boiling fractions were only reacted at the higher temperatures.

The explanation for the wide range of reactivity is changes in substitution of heterocycles and in more refractory benzologs in higher-boiling fractions (Speight, 1991). The relative ease of removal of the two types of nitrogen, therefore, depends on feed and reactor characteristics.

Thermal reactions also occur during hydrotreating at ca. 400°C (750°F) and the formation of C_1-C_3 hydrocarbons is independent of the presence of a catalyst (Khorasheh et al., 1989). However, the yields of C_4 and C_5 are enhanced by the presence of a catalyst, consistent with hydrogenation of aromatics followed by cracking and ring-opening reactions.

The occurrence of reactions that lead to the formation of hydrocarbon species is, on the surface, the preferential way to proceed, but there are potential drawbacks. The generation of hydrocarbon liquids in the reaction mix may be analogous to the addition of a hydrocarbon liquid during the deasphalting procedure. This leads to the precipitation of an asphaltene fraction, which consists mainly of polar species and high-molecular-weight species. Similarly, the enhancement of the paraffinicity of the liquid phase at the same time that polar species remain in solution can lead to sediment, or sludge, formation during the reaction and, hence, the onset of incompatibility.

4.0 References

Benson, S.W. and Shaw, R. 1970. Organic Peroxides. Volume I.
    Wiley Interscience, New York. Chapter 2.

Bridgewater, A.J., and Sexton, M.D. 1978. J. Chem. Soc. Perkin
    Trans. 530.

Chmielowiec, J., Fischer, P., and Pyburn, C.M. 1987. Fuel 66:
    1358.

Cooney, J.V., Beal, E.J., and Hazlett, R.N. 1984. Liquid Fuels
    Technol. 2: 395.

Coordinating Research Council. 1979. Literature Survey on the
    Thermal Oxidation Stability of Jet Fuel. CRC Report No. 509.
    CRC Inc., Atlanta, Georgia.

Coordinating Research Council. 1988. Determination of the
    Hydroperoxide Potential of Jet Fuels. CRC Report No. 559,
    CRC Inc., Atlanta, Georgia.

Curci, R., Giovine, A., and Modena, G. 1966. Tetrahedron. 22:
    1235.

Danehy, J.P. 1968. Mechanism of Reactions of Sulfur Compounds.
    Volume 2. P. 69.

Denisov, E.T. 1974. Liquid Phase Reaction Rate Constants.
    IFI/Plenum, New York.

Dorbon, M., Ignatiadis, I., Schmitter, J.M., Arpino, P.,
    Guiochon, G., Toulhoat, H., and Huc, H. 1984. Fuel. 63: 565.

Fava, A., Reichenbach, G., and Peron, U. 1967. J. Am. Chem. Soc.
    89: 6696.

Field, L. 1977. Organic Chemistry of Sulfur Compounds. IFI/Plenum,
    New York. PP. 325-352.

Furimsky, E., Ranganathan, R., and Parsons, B.I. 1978. Fuel. 57:
    494.

Hardy, D.R., and Wechter, M.A. 1990. Energy Fuels. 4: 270.

Hartough, H.D. and S.L. Meisel. 1954. Compounds with Condensed
    Thiophene Rings. Wiley-Interscience, New York.

Hartough, H.D. 1964. Advances in Petroleum Chemistry and Refining
    Wiley-Interscience, New York.

Hazlett, R.N., Schreifels, J.A., Stalick, W.F., Morris, R.E.,
    Mushrush, G.W. 1991. Energy Fuels. 5:269.

Henbest, H. B., and Khan, K. A. 1968. Chem. Commun. 17: 1036.

Hiatt, R. R., and Strachan, W. M. J. 1963. J. Org. Chem. 28,
    1893.

PETROLEUM PRODUCTS: INSTABILITY AND INCOMPATIBILITY

Howard, J.A., The Chemistry of Functional Groups, Peroxides, Wiley-Interscience, New York, 1983.

Hucknall, D.J. 1974. Selective Oxidation of Hydrocarbons. Academic Press, New York.

Joule, J.A. and Smith, G.F. 1979. Heterocyclic Chemistry. Van Nostrand Reinhold, New York.

Kharasch, N. (Editor). 1961. Organic Sulfur Compounds. Volume I. New York. Pergamon Press.

Kharasch, N. and Meyers, C.Y. (Editors). 1966. The Chemistry of Organic Sulfur Compounds. Volume 2. New York, Pergamon Press.

Khorasheh, F., Rangwala, H., Gray, M.R. and Dalla Lana, I.G. 1989. Energy Fuel. 3: 716.

Kong, R., Lee, M.L., Iwao, M.Y., Tominaga, Pretaap, R.,Thompson, R.D., and Castle, R.N. 1984. Fuel 63: 703.

Liotta, R., and Hoff, W.S. 1980. J. Org. Chem.45: 2887.

Mayo, F.R. 1972. The Chemistry of Fuel Deposits and Their Precursors. Final Report N00019-72-C-10161. Naval Air Systems Command., Washington, DC.

Migita, T., Kosugi, M., Takayama, K., Nakagawa, Y., 1973 Tetrahedron 29: 51.

Modena, G., and Todesco, P.E. 1962 J. Chem. Soc. 4920.

Mosher, H.S. 1960 J. Am. Chem. Soc. 82: 4537.

Mushrush, G.W. and Hazlett, R. N. 1985 J. Org. Chem. 50: 2387.

Mushrush, G.W., Hazlett, R.N., Hardy, D.R., and Watkins, J.M. 1987. Ind. Eng. Chem. Res. 26: 662.

Mushrush, G.W., Beal, E.J., Pellenbarg, R.E., Hazlett, R.N., Eaton, H.R., and Hardy, D.R. 1994. Energy Fuels. 8: 851.

Mushrush, G.W., Pellenburg, R.E., Hazlett, R.N., Morris, R.E., Hardy, D.R. 1991. Fuel Science & Tech. 9:1137.

Nixon, A.C. 1962. Autoxidation and Antioxidants in Petroleum. Wiley Interscience, New York.

Oae, S. 1977. Organic Chemistry of Sulfur Compounds, IFI/Plenum, New York.

Offenhauer, R.D., Brennan, J.A., and Miller, R.C. 1956. Ind. Eng. Chem. 49: 1265.

Ohno, A., Kito, N., and Ohnmish, Y. 1971. Bull. Chem. Soc. Jpn. 44:463.

Oswald, A.A., Noel, F.J.. 1961. J. Chem. Eng. Data. 6: 294.

Rahman, A., and Williams, A. 1970. J. Chem. Soc. B: 1391.

Reid, E.E. 1958. Organic Chemistry of Bivalent Sulfer. Volume I. Chemical Publishing Co., New York.

200

Speight, J. G. 1991, The Chemistry and Technology of Petroleum. 2nd Edition. Marcel Dekker Inc., New York.

Taylor, W. F. 1976. Ind. Eng. Chem. Prod. Res. Dev. 7: 64.

Taylor, W. F., and Wallace, T. J. 1967. Ind. Eng. Chem. Prod. Res. Dev. 6: 258.

Thompson, R. B., Druge, L. W., and Chenicek, J. A. 1949. Ind. Eng. Chem. 41: 2715.

Trytten, L. C., Gray, M. R., and Sandford, E. C. 1990. Ind. Eng. Chem. Res. 29: 725.

Turney, T. A. 1965. Oxidation Mechanisms. Butterworths, London.

Van Swet, H., and Kooyman, E. 1968. Recl. Trav. Chim. Pays-Bas. 87:45.

Walters, E. L., Minor, H. B., and Yabroff, D. L. 1949. Ind. Eng. Chem. 41: 1723.

Watkins, J. M., Mushrush, G. W., Hazlett, R. N., and Beal, E. J. 1989. Energy Fuels. 3: 231.

CHAPTER 7: FUEL PROPERTIES AND INSTABILITY/INCOMPATIBILITY

1.0 Introduction
The chemistry and physics of incompatibility can, to some extent, be elucidated (Chapter 6) (Wallace, 1969) but many unknowns remain. In addition to the chemical aspects, there are also such aspects as the attractive force differences, e.g.

(1) specific interactions between like/unlike molecules (e.g. hydrogen bonding and electron donor-acceptor phenomena) that are also effective;

(2) field interactions such as dispersion forces and dipole-dipole interactions; and

(3) any effects imposed on the system by the size and shape of the interacting molecular species.

Such interactions are not always easy to define, and thus, the measurement of instability and incompatibility has, by convention, involved a gravimetric method whereby the sediment formed over a given time period was determined. For example, visual observations, solubility tests, as well as other standard tests

Table 7.1: Identification and measurement of instability.

- Compatibility
    Visual observations
    Solubility tests
    Hot filtration sediment (HFS)

- Storage Stability
    Existent gum (ASTM D-381)
    Potential gum (ASTM D-873)
    Accelerated storage test

- Thermal Stability
    JFTOT (ASTM D-3241)
    Exxon advanced fuel unit

(Table 7.1) have evolved as a means of broaching the issue of instability/incompatibility.

However, such methods are often considered to be after-the-fact methods insofar as they did not offer much in the way of predictability. In refinery processes, predictability is not just a luxury, it is a necessity. The same principle must be applied to the measurement of instability and incompatibility. Therefore, methods are continually being sought to aid in achieving this goal.

It is the purpose of this chapter to document some of the more prominent methods used for determining instability and incompatibility. No preference will be shown, for any individual method. It is the choice of the individual experimentalist to choose a method on the basis of the type of fuel, the immediate needs, and the projected utilization of the data. As elsewhere, it is the use of the data that often detracts from an otherwise sound method.

In addition to the gravimetric methods, there have been many attempts to use crude oil and/or product characteristics and their relation to the sludge and deposit formation tendencies. In some cases, a modicum of predictability is the outcome, but in many cases, the data appear as **preferred ranges** and are subject to personal interpretation (Speight, 1991). Therefore, caution is advised.

2.0 Instability/Incompatibility of Liquid Fuels and Products

The instability/incompatibility of crude oil and of crude oil products is manifested in the formation of sludge, sediment, and general decoloration. The decoloration usually takes the form of darkening of the liquid.

Sludge (or sediment), after formation, takes one of the following forms: (1) material dissolved in the liquid; (2) precipitated material; and (3) material emulsified in the liquid.

Under favorable conditions, sludge or sediment will dissolve in the crude oil or product with the potential of increasing the viscosity. Sludge/sediment that does not dissolve in crude oil may either settle at the bottom of the storage tanks or remain in the crude oil as emulsions. In most cases the smaller part of the sludge/sediment will settle satisfactorily, and the larger part will stay in the crude oil as emulsions. In any case there is a need to break up the emulsion, whether it is a water in oil emulsion or whether it is the sludge itself, which has to be separated into the oily phase and the aqueous phase. The oily phase can then be processed with the crude oil and the aqueous phase can be drained out of the system.

The phase separation can be accomplished by either the use of suitable surface active agents, allowing for sufficient settling time, or by use of a high voltage electric field for breaking such emulsions after admixing water at a rate of about 5% and at a temperature of about 100°C (212°F)

The breaking of emulsions, whether they are crude oil-sludge emulsions or crude oil-water emulsions, or the breaking of the sludge itself into its oily and inorganic components, is of a major importance from an operational as well as a commercial viewpoint.

With some crude oils, especially the heavy and the viscous oils, phase separation difficulties often arise. Also, some crude oil emulsions may be stabilized by naturally occurring substances in the crude oil. Many of these polar particles accumulate at the oil-water interphase, the polar groups being directed toward the water and the hydrocarbon groups toward the oil. A stable interfacial skin may be so formed; particles of clay or similar impurities, as well as wax crystals present in the oil may be embedded in this skin and make the emulsion very difficult to break.

Chemical and electrical desludging/dewatering operations, often combined with chemical additives, have to be used for breaking such emulsions. Each emulsion has its own structure and characteristics: water in oil emulsions, where the oil is the major component, or oil in water emulsions, where the water is the major component. The chemical and physical nature of the emulsion components plays a major role in how susceptible the components are to the various surface active agents used for breaking them.

Therefore, appropriate emulsion breaking agents have to be chosen very carefully, usually with the help of previous laboratory evaluations. Water or oil soluble demulsifiers, the latter being often nonionic surface active alkylene oxide adducts, are used for this purpose. However, as was stated in the foregoing, the most suitable demulsifier has to be chosen for each case from a large number of such substances in the market, by prior laboratory evaluation.

Some heavy crude oils have been found to be especially

susceptible to sludge formation. In most cases an increase of viscosities and pour points was observed. In addition, there is a drop in the initial asphaltene content of the treated samples, but during storage the asphaltene content increases more in the treated than in samples.

The demulsification process accelerates the formation of degradation products during storage of some crude oils.

The aromatic character of crude oil is favorable to a dissolution of degradation products in this crude oil during the storage period.

The conclusions drawn from several studies are instructive in respect to the sludge forming tendencies of the various crude oil types. In some cases, the sludge separates, while in others, a part or all of the sludge dissolves in the crude, depending on the chemical characteristics of the crude oil and on the chemical composition of the sludge. The dissolution of the sludge in the crude oil is accompanied by an increase of the crude oil viscosity.

3.0 Factors Influencing Instability/Incompatibility

A number of experimental methods are available for estimation of the factors that influence instability/incompatibility. These factors have been explored and attempts made to estimate the character of the fuel or product, and results have been varied.

3.1 Elemental Analysis

The ultimate analysis (elemental composition) of petroleum and its products is not reported to the same extent as for coal (Berkowitz, 1979; Hessley et al., 1986; Speight, 1994). Nevertheless, there are ASTM procedures (ASTM, 1993) for the ultimate analysis of petroleum and petroleum products, but many such methods may have been designed for other materials.

For example, carbon can be determined by the method designated for coal and coke (ASTM D-3178) or by the method designated for municipal solid waste (ASTM E-777). There are also methods designated for hydrogen (ASTM D-1018, D-3178, D-3343, D-3701, and E-777), nitrogen (ASTM D-3179, D-3228, D-3431, E-148, E-258, and E-778), oxygen (ASTM E-385) and sulfur (ASTM D-124, D-1266, D-1552, D-1757, D-2662, D-3177, D-4045 and D-4294). Many of these methods have been subject to modification by various users because of the nature of the material under investigation. Some of these modifications may have also led to the development of proprietary in-house methods.

Of the data that are available (Speight, 1991), the proportions of the elements in petroleum, unlike coal (Speight, 1994), vary only slightly over narrow limits:

Carbon 83.0 - 87.0%
Hydrogen 10.0 - 14.0%
Nitrogen 0.10 - 2.0%
Oxygen 0.05 - 1.5%
Sulfur 0.05 - 6.0%

Yet there is a wide variation in physical properties from the lighter, more mobile crude oils, at one extreme, to the heavier asphaltic crude oils at the other extreme (Figure 7.1). The majority of the more aromatic species and the heteroatoms occur in the higher boiling fractions of feedstocks (Figure 7.2). The heavier feedstocks are relatively rich in these higher boiling fractions.

In terms of the incompatibility of petroleum, liquid fuels, and products, the heteroatom content probably represents the greatest influence.

Indeed, the very nature of the distillation process by which residua are produced, i.e., removal of distillate without thermal decomposition, dictates that the majority of the heteroatoms, which are predominantly of the higher molecular weight fractions, be concentrated in the higher boiling products and the residuum (Speight, 1981). Thus, the inherent nature of the crude oil and the means by which it is refined can seriously influence the stability and incompatibility of the products.

In fact, the sulfur and nitrogen content of crude oil is an important parameter in respect to the processing methods that have to be used in order to produce fuels of specification sulfur concentrations. There could well be a relation between nitrogen and sulfur content and crude oil (or product) stability; higher nitrogen and sulfur crude oils are suspect of higher sludge forming tendencies.

207

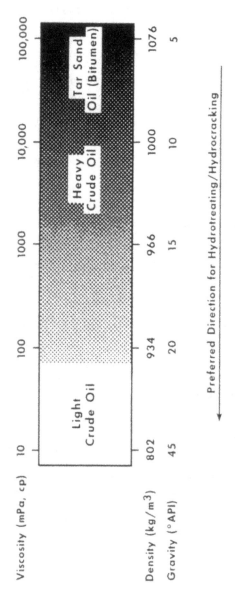

Figure 7.1: Schematic Representation of the Range of Crude Oil Viscosities, Densities, and API Gravities.

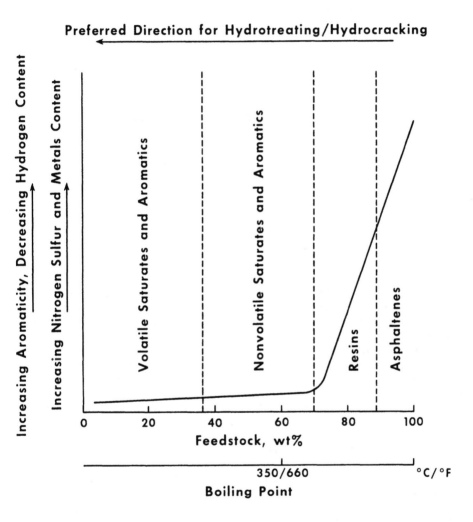

Figure 7.2: Schematic Representation of the Influence of Boiling Point on Aromatics, Nitrogen, Sulfur, and Metals Content.

Table 7.2: Crude petroleum is a mixture of compounds that can be separated into different generic boiling fractions.

Fraction	Boiling Range °C	°F
Light naphtha	-1-150	30-300
Gasoline	-1-180	30-355
Heavy naphtha	150-205	300-400
Kerosene	205-260	400-500
Stove oil	205-290	400-550
Light gas oil	260-315	400-600
Heavy gas oil	315-425	600-800
Lubricating oil	>400	>750
Vacuum gas oil	425-600	800-1050
Residuum	>600	>1050

3.2 Density/Specific Gravity

In the early years of the petroleum industry, density and specific gravity (ASTM D-287, D-1298, D-941, D1217 and D-1555) were the principal specifications for feedstocks and refinery products. They were used to give an estimate of the most desirable product, i.e. kerosene, in crude oil. Specific gravity, which closely (but not exactly) approximates density, is the most common term at present but suffers from the disadvantage that only a very narrow range of values applies to a wide range of feedstocks (Speight, 1991).

The American Petroleum Institute (API) devised a wider scale that is based on the inverse of the specific gravity scale:

$$\text{Degrees API} = (141.5/d\ 60°/60°F) - 131.5$$

As has already been noted, it is possible to recognize certain trends between the API gravity and, for example, sulfur content (Figure 7.3a), nitrogen content (Figure 7.3b), viscosity (Figure 7.4a), carbon residue (Figure 7.4b), and asphaltene or asphaltics (asphaltene + resin) content (Figure 7.5). Hydrocracked products, containing very little sulfur, are at the higher end (>30°) of the

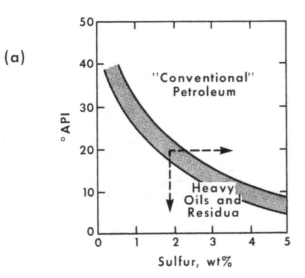

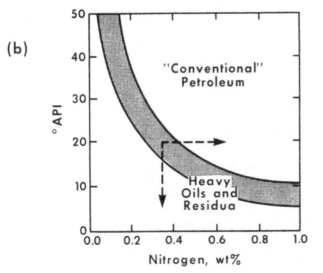

Figure 7.3: Schematic Representation of (a) Sulfur and (b) Nitrogen Contents of Various Feedstocks.

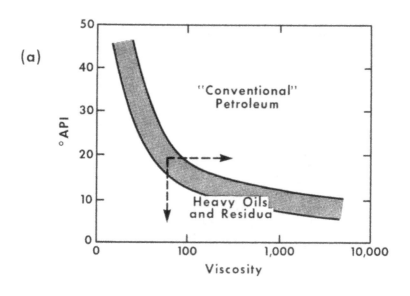

(a)

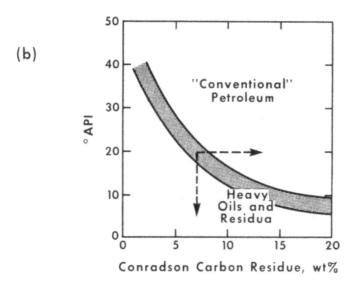

(b)

Figure 7.4: Schematic Representation of (a) the Viscosities and (b) the Carbon Residues for Various Feedstocks.

212

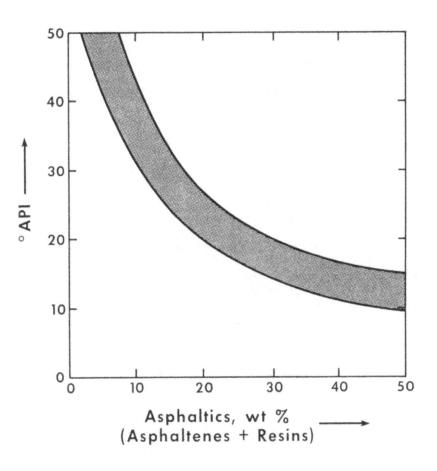

Figure 7.5: Relationship of API Gravity to Asphaltenes Plus Resins Content.

API scale.

There is no direct relation between the density or specific gravity of crude oils and their sludge forming tendencies, but crude oil having a higher density (thus, a lower API gravity) is generally more susceptible to sludge formation, presumably because of the higher content of the polar/asphaltic constituents.

3.3 Volatility

Petroleum can be subdivided by distillation into a variety of fractions of different "cut points" (Table 7.2). In fact, this was, and still is, the method by which petroleum feedstocks were evaluated as being suitable for various refinery options. Indeed, volatility is one of the major tests for petroleum products and it is inevitable that all hydrocracked products will, at some stage of their history, be tested for volatility characteristics.

As an early part of characterization studies, a correlation was observed between the quality of petroleum products and their hydrogen content (Figure 7.6) since gasolines, kerosenes, diesel fuels, and lubricating oils are made up of hydrocarbon constituents containing high proportions of hydrogen. Thus, it is not surprising that tests to determine the volatility of petroleum and petroleum products were among the first to be defined.

The vaporizing tendencies of liquid fuels are the basis for the general characterization of petroleum fuels such as liquefied petroleum gas, natural gasoline, motor and aviation gasolines, naphthas, kerosene, gas oils, diesel fuels, and fuel oils (ASTM D-2715). The flash point is the temperature at which the product must be heated under specified conditions to give off sufficient vapor to form a mixture with air that can be ignited momentarily by a flame (ASTM D-56, D-92, D-93). The fire point is the temperature at which petroleum or a petroleum product must be heated to burn continuously when a mixture of the vapor and air is ignited by a flame (ASTM D-92). The vapor pressure of petroleum and its constituent fractions is the force that must be exerted on the liquid to prevent it from vaporizing further (ASTM D-323).

Distillation involves the general procedure of vaporizing a liquid either at atmospheric pressure (ASTM D-86, D-216, D-285, D-

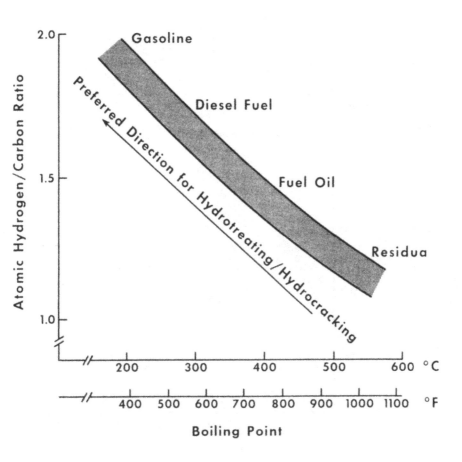

Figure 7.6: General Representation of the Relationship of Boiling Point of Petroleum Products to Atomic Hydrogen/Carbon Ratio.

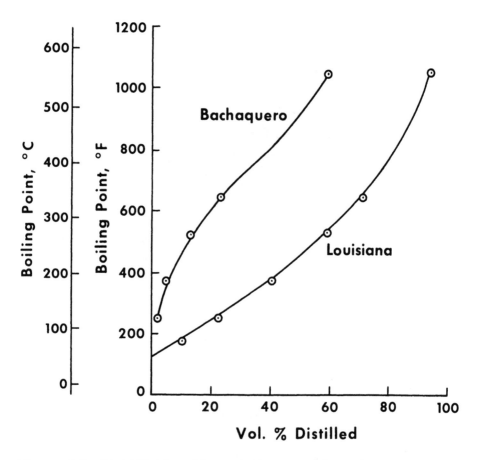

Figure 7.7: Distillation "Curves" for Two Selected Crude Oils.

Table 7.3: Distillation specifications (U.S. Bureau of Mines
Routine Method) for a light(Leduc) crude oil.

Cut at		Wt.	Comp.	
°C	°F	%	%	°API

(a)
Stage 1: Distillation at atmospheric pressure

50	122	2.8	2.8	84.2
100	212	5.1	11.2	67.0
150	302	6.2	24.2	54.4
200	392	5.0	35.1	46.7
250	482	5.1	45.2	40.9
275	527	6.1	51.3	38.4

Stage 2: Distillation continued reduced pressure

200	392	5.2	56.5	35.2
225	437	4.8	61.3	33.4
250	482	5.0	66.3	31.1
275	527	4.7	71.0	28.9
300	572	4.8	75.8	27.3
Residuum		20.2	96.0	18.1

Carbon residue of residuum: 6.0%

447 and D-2892) or at reduced pressures (ASTM D-1160; ASTM D-196) and the data are reported in terms of one, or more, of the following items: initial boiling point; distillation temperature; end point; dry point; percent distillate; percent residue; percent total recovery, i.e., sum of the liquids and residue; percent distillation loss; and percent evaporated at a specific thermometer reading or other distillation temperatures (Table 7.3; Figure 7.7). One of the methods (ASTM D-285) was discontinued in 1987, but as often occurs when discontinuation of a method is announced, many laboratories will continue to use the technique for some time to come.

There are some limitations to the routine distillation tests. For example, although heavy crude oils do contain volatile constituents, it is not always advisable to use distillation for identification of these volatile constituents. Thermal

217

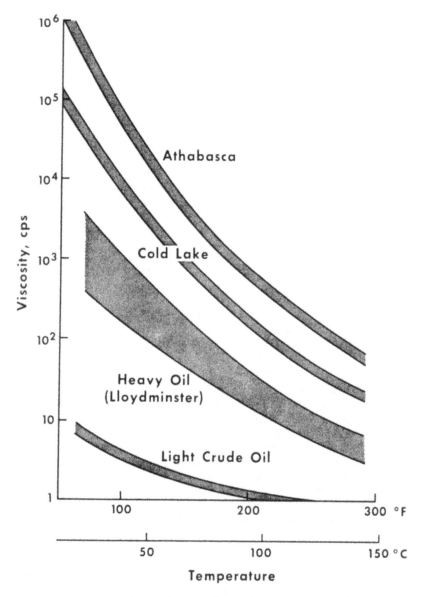

Figure 7.8: Relationship of Viscosity to Crude Oil.

218

decomposition of the constituents of petroleum is known to occur at approximately 350°C (660°F), but thermal decomposition of the constituents of the heavier, but immature, crude oils has been known to commence at temperatures as low as 200°C (390°F). Thus, thermal alteration of the constituents and erroneous identification of the decomposition products as "natural" constituents are always possible.

On the other hand, the limitations to the use of distillation as an identification technique might be economical, and detailed fractionation of the sample might also be of secondary importance. There have been attempts to combat these limitations, but it must be recognized that the general shape of a one-plate distillation curve is often adequate for making engineering calculations, correlating with other physical properties, and predicting the product slate (Nelson, 1958).

However, a low-resolution, temperature-programmed gas chromatographic analysis has been developed to simulate the time-consuming true boiling point distillation (ASTM D-2887). The method relies on the general observation that hydrocarbons are eluted from a nonpolar adsorbent in the order of their boiling points. The regularity of the elution order of the hydrocarbon components allows the retention times to be equated to distillation temperatures (Green et al., 1964), and the term "simulated distillation by gas chromatography" (or "simdis") is used throughout the industry to refer to this technique.

There is another method that is increasing in popularity for application to a variety of feedstocks, and that is the method commonly known as "simulated distillation" (ASTM D-2887). The method has been well researched in terms of method development and application (Hickerson, 1975; Green, 1976; Stuckey, 1978; Vercier and Mouton, 1979; Thomas et al. 1983; Romanowski and Thomas, 1985; MacAllister and DeRuiter, 1987; Schwartz et al., 1987).

The benefits of the technique include good comparisons with other ASTM distillation data as well as the application to higher boiling fractions of petroleum. In fact, data output includes the provision of the corresponding Engler profile (ASTM D-86) as well as the prediction of other properties such as vapor pressure and flash point (DeBruine and Ellison, 1973). When it is necessary to monitor product properties, as is often the case during hydrocracking operations, such data provide a valuable aid to process control and on-line product testing.

Simulated distillation by gas chromatography is often applied in the petroleum industry to obtain true boiling point data for distillates and crude oils (Butler, 1979). Two standardized methods, ASTM D-2887 and ASTM D-3710, are available for the boiling point determination of petroleum fractions and gasoline, respectively. The ASTM D-2887 method utilizes nonpolar, packed gas chromatographic columns in conjunction with flame ionization detection. The upper limit of the boiling range covered by this method is to approximately 540°C (1000°F) atmospheric equivalent boiling point. Recent efforts in which high temperature gas chromatography was used have focused on extending the scope of the ASTM D-2887 methods for higher boiling petroleum materials to 800°C (1470°F) atmospheric equivalent boiling point (Schwartz et al., 1987).

Heavier crude oils, yielding higher residue rates, tend to form more sludge during storage in comparison with light crude oils.

Table 7.4: Viscosity ranges for various petroleum products.

Product	Viscosity (poise)
Paraffins	0.4×10^{-2}
Cycloparaffins	0.6×10^{-2}
Alkylbenzenes	0.8×10^{-2}
Gasoline	0.6×10^{-2}
Kerosene	2.0×10^{-2}
Lubricating oil	$10^{-2} - 10^{3}$
Residua	$10^{3} - 10^{7}$

3.4 Viscosity

The viscosity of a feedstock varies with the origin and type of the crude oil and also with the nature of the chemical constituents, particularly the polar functions, where intermolecular interactions can occur. For example, there is a gradation of viscosity among conventional crude oil, heavy oil, extra heavy oil, and bitumen (Figures 7.8 and 7.9).

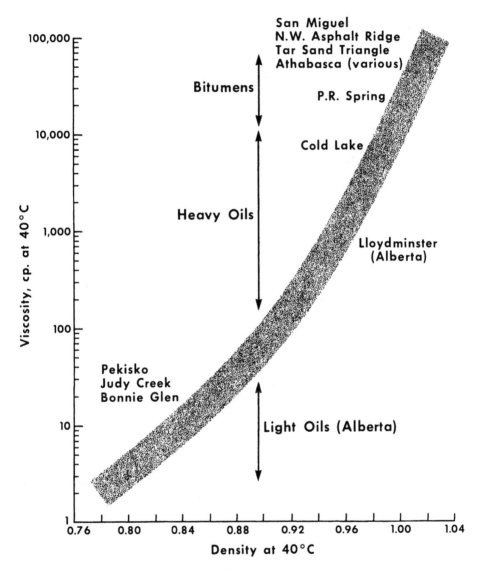

Figure 7.9: Schematic Representation of the Increase in Viscosity With the "Heaviness" of the Feedstock.

221

There are several methods for the determination of viscosity, and the units of viscosity vary depending upon the method (ASTM D-445, D-88, D-2161, D-341 and D-2270). Viscosity data may also be reported as Saybolt universal seconds (SUS) or as Saybolt furol seconds (SFS), which by means of conversion factors, can be reported as stokes. It is possible to interconvert the several viscosity scales (Speight, 1991), especially Saybolt to kinematic viscosity (ASTM D-2161).

However, viscosity is often expressed in units of "stokes" or in units of "poises" (poises = stokes x specific gravity), and the viscosity of crude oils may vary from several stokes for hydrocracked liquids and the lighter crude oils to several thousand stokes for heavier crude oils, and even to several million stokes for residua and bitumens (Figures 7.8 and 7.9). There is also a considerable variation in the viscosities of individual petroleum fractions and petroleum products (Table 7.4).

Viscosities indicate fluidity properties and consistencies at given temperatures. Heavier crude oils, i.e. crude oils of lower API gravities, usually have higher viscosities. Increases of viscosities during storage indicate either an evaporation of volatile components or formation of degradation products dissolving in the crude oil.

3.5 Asphaltene Content and Carbon Residue

The asphaltene fraction (IP-143) of feedstocks is particularly important because, as the proportion of this fraction increases, there is a concomitant increase in thermal coke carbon residue (Figure 7.10a) and an increase in hydrogen demand as well as catalyst deactivation (Figure 7.10b). Thus, the result of a carbon residue test can be indicative of the amount of asphaltenes in a feedstock (Figure 7.11) (Speight, 1987) and, therefore, indicative of the potential hydrogen requirements as well as the potential for coke deposition on the catalyst. This latter phenomenon is of particular interest in terms of the compatibility/incompatibility of the coke precursors.

Polar materials, i.e., resins, that remain in the deasphaltened oil are also capable of forming carbon residues (Figure 7.11) or coke deposits on catalyst surfaces, thereby diminishing the effectiveness of the catalyst.

All thermal processes involving heavy feedstocks produce coke

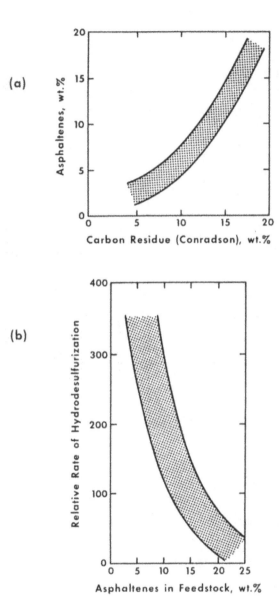

(a)

Asphaltenes, wt.%

Carbon Residue (Conradson), wt.%

(b)

Relative Rate of Hydrodesulfurization

Asphaltenes in Feedstock, wt.%

Figure 7.10: Schematic Representation of (a) Carbon Residue and (b) Relative Rate of Catalyst Deactivation and Hydrogen Consumption to Asphaltene Content.

223

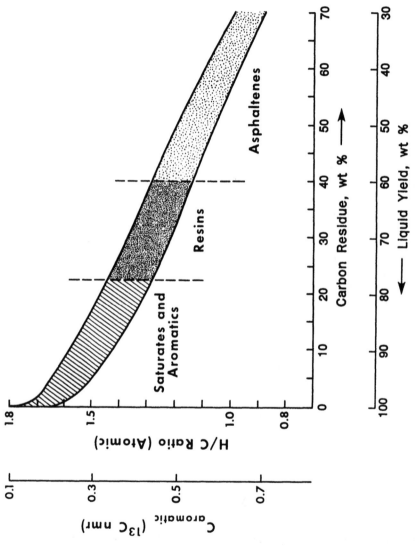

Figure 7.11: Aromaticity and Carbon Residue for Various Fractions.

(Nelson, 1976; Speight, 1991). Therefore, tests have been designed to determine the amount of thermal coke (carbon residue) that is likely to be produced by a feedstock (Nelson, 1964, 1974a, 1978). There are two older, well-used methods for determining the carbon residue: the Conradson method (ASTM D-189) and the Ramsbottom method (ASTM D-524). Both are equally applicable to the heavier feedstocks, and it is possible to interconnect the data (Speight, 1991). Recently, a newer method has been accepted by the ASTM (ASTM D-4530) that was originally developed as a thermogravimetric method and agreement between the data from the three methods is good, thereby making it possible to interrelate all of the data from carbon residue tests (Long and Speight, 1989).

Determination of the carbon residues of lower boiling fractions of petroleum, of the lighter petroleums, or of hydrocracked products may be difficult because of the high volatility of the material and the low yield of the carbon residue. To overcome the extremely small values of carbon residue when the tests are applied to such materials, it is customary to distill such products to 10% residual oil and determine the carbon residue thereof.

There are general relationships (Nelson, 1974b; Speight, 1981, 1989) between the carbon residue of a feedstock and parameters such as the asphaltene content and the API gravity that extend the relationship of the carbon residue to several other feedstock parameters such as sulfur content, nitrogen content, metals content, and viscosity (Figure 7.12). Recent work has focused on the precise relationship between carbon residue and hydrogen content, H/C atomic ratio, nitrogen content, and sulfur content (Figure 7.13), thereby providing more precise information about the anticipated behavior of a variety of different feedstocks in thermal processes (Roberts, 1989).

The effect of asphaltenes, micellar structure, and state of peptization also merit some attention. The peptization state of asphaltenes is higher in the more naphthenic/aromatic crude oils because of higher solvency of naphthenes and aromatics over paraffinic constituents. This phenomenon also acts in favor of the dissolution of any sludge that may form thereby tending to decrease sludge deposition. However, an increase in crude oil often accompanies sludge dissolution.

The higher the asphaltenes and the carbon residue, the higher

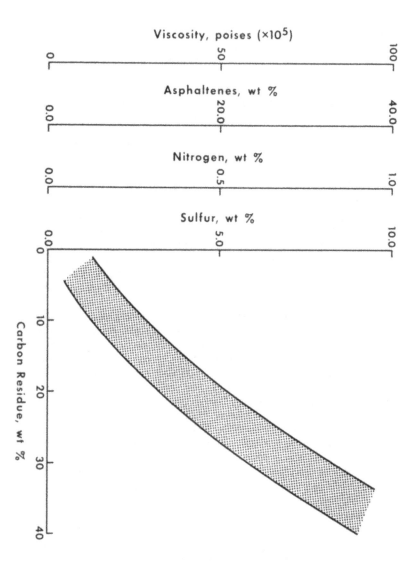

Figure 7.12: Generalized Relationship of Several Crude Oil Properties.

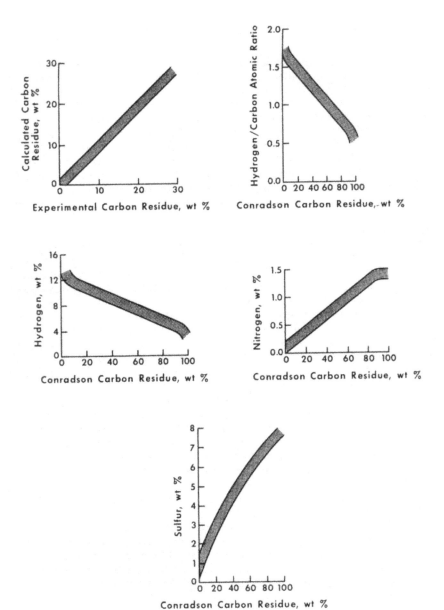

Figure 7.13: Relationship of Various Properties to Carbon Residue.

will be the tendency of the crude oil to form sludge, especially when blended with other noncompatible stocks.

3.6 Pour Point

Pour points define the cold properties of crude oils and petroleum products, i.e., the minimal temperature at which they still retain their fluidity. Pour points also indicate the characteristics of crude oils: the higher the pour point, the more paraffinic is the oil, and vice versa. Higher pour point crude oils are more waxy and, therefore, tend to form wax-like materials that enhance sludge formation.

3.7 Total Acidity

Total acidities consist of strong organic acidities and weak organic acidities. Inorganic acidity is not expected to be present in crude oils. Values above 0.15 mg potassium hydroxide per gram are considered to be significantly high. Crude oils of higher acidities may exhibit a tendency toward instability.

The main agents for inparting acidity in crude oils are naphthenic acids and hydrogen sulfides. These are sometimes present in the crude oil originally in small and varying concentrations. Normally, the total acidity of crude oils is in the range of 0.1 to 0.5 mg potassium hydroxide per gram, although higher values are not exceptional.

Free hydrogen sulfide is often present in crude oils, a concentration of up to 10 ppm being acceptable in spite of its toxic nature. However, higher concentrations of hydrogen sulfide are sometimes present; 20 ppm would pose serious safety hazards. Additional amounts of hydrogen sulfide can form during the crude oil processing, when hydrogen reacts with some organic sulfur compounds, converting them to hydrogen sulfide. In this case it is referred to as potential hydrogen sulfide, as opposed to free hydrogen sulfide.

Acidity can also form by bacterial action insofar as some species of aerobic bacteria can produce organic acids from organic nutrients. On the other hand, anaerobic sulfate-reducing bacteria can generate hydrogen sulfide, which in turn, can be converted to sulfuric acid (by bacterial action).

Table 7.5:   Ranges of principal trace elements found in petroleum.

Element	Range in petroleum ppm		
Cu	0.2	-	12.0
Ca	1.0	-	2.5
Mg	1.0	-	2.5
Ba	0.001	-	0.1
Sr	0.001	-	0.1
Zn	0.5	-	1.0
Hg	0.03	-	0.1
Ce	0.001	-	0.6
B	0.001	-	0.1
Al	0.5	-	1.0
Ga	0.001	-	0.1
Ti	0.001	-	0.4
Zr	0.001	-	0.4
Si	0.1	-	5.0
Sn	0.1	-	0.3
Pb	0.001	-	0.2
V	5.0	-	1500
Fe	0.04	-	120
Co	0.001	-	12
Ni	3.0	-	120

3.8 Metals (Ash) Content

The majority of crude oils contain metallic constituents, particularly in the heavier feedstocks. These constituents, of which nickel and vanadium are the principal metals, are very influential in regard to feedstock behavior in processing operations (Table 7.5).

The metal (inorganic) constituents of petroleum or a liquid fuel are either present in the crude oil originally or are picked up by the crude oil during storage and handling. The former are mostly metallic substances like vanadium, nickel, sodium, iron, and silica; the latter may be contaminants like sand, dust and corrosion products.

Incompatibility, leading to deposition of the metals (in any form) onto the catalyst, leads to catalyst deactivation whether it be by physical blockage of the pores or by destruction of reactive sites. In the present context, the metals must first be removed if erroneously high carbon residue data are to be avoided. Alternatively, the metals can be estimated as ash by complete burning of the coke after carbon residue determination.

There are many methods proposed by the ASTM for metals detection in petroleum and petroleum products. One particular method (ASTM D-3605) is the more general of these, and the latest ASTM volumes should be consulted for more detail.

Metals content above 200 ppm is considered to be significant, but the variations are very large. The higher the ash content the higher is the tendency of the crude oil to form sludge or sediment.

3.9 Water Content

Water content, salt content and bottom sediment/water indicate the concentrations of aqueous contaminants, present in the crude either originally or after being picked up during handling and storage. Water and salt content of crude oils produced at the field can be very high, sometimes forming its major part. The salty water is usually separated at the field, by settling and draining, and surface active agents electrical emulsion breakers (desalters) are sometimes employed. Water content below 0.5%, salt content up to 20 pounds per 1000 barrels, and bottom sediment/water up to 0.5% are considered to be satisfactory.

The higher the water content and the BS&W, the higher are the sludge and deposit formation rates that can be expected in the stored crude oil.

3.10 Characterization Indices

The **characterization factor** indicates the chemical character of the crude oil and has been used to indicate whether a crude oil is paraffinic in nature or is a naphthenic/aromatic crude oil.

The characterization factor (sometimes referred to as the Watson characterization factor) is given by the relation:

$$K = T_b^{1/3}/d$$

where T_b is the average boiling point, degrees Rankine (°F + 460), and d is the specific gravity at 15.6°C (60°F).

230

The characterization factor was originally devised to illustrate the characteristics of various feedstocks. Highly paraffinic oils have K = 12.5-13.0, whilst naphthenic oils have K = 10.5-12.5. In addition, if the characterization factor is above 12, the liquid fuel or product might, because of its paraffinic nature, be expected to form waxy deposits during storage.

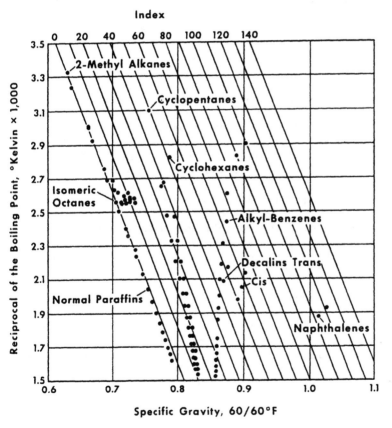

Figure 7.14: Reference Data for the correlation Index.

PETROLEUM PRODUCTS: INSTABILITY AND INCOMPATIBILITY

The **viscosity-gravity constant** (vgc) was one of the early indices proposed to classify petroleum on the basis of composition. It is particularly valuable for indicating a predominantly paraffinic or naphthenic composition.

The constant is based on the differences between the density and specific gravity for the various hydrocarbon species:

$$vgc = [10d - 1.0752 \log(v - 380)]/[10 - \log (v - 38)]$$

where d is the specific gravity and v is the Saybolt viscosity at 38°C (100°F). For viscous crude oils (and viscous products) where the viscosity is difficult to measure at low temperatures, the viscosity at 99°C (210°F) can be used:

$$vgc = [d - 0.24 - 0.022 \log(v - 35.5)]/0.755$$

In both cases, the lower the index number, the more paraffinic the sample. For example, a paraffinic sample may have a vgc of the order of 0.840, whilst the corresponding naphthenic sample may have an index of the order of 0.876.

The obvious disadvantage is the closeness of the indices, almost analogous to comparing crude oil character by specific gravity only where most crude oils fall into the range d = 0.800-1.000. The API gravity expanded this scale from 5 to 60 thereby adding more meaning to the use of specific gravity data.

In a similar manner, the **correlation index** which is based on a plot of the specific gravity (d) versus the reciprocal of the boiling point (K) in °K (°K = degrees Kelvin = °C + 273) for pure hydrocarbons, adds another dimension to the numbers:

$$Correlation\ Index\ (CI) = 473.7d - 456.8 + 48,640/K$$

In the case of a petroleum fraction, K is the average boiling point determined by the standard distillation method.

The line described by the constants of the individual members of the normal paraffin series is given a value of CI = 0 and a parallel line passing through the point for benzene is given a value of CI = 100 (Figure 7.14). Values between 0 and 15 indicate a predominance of paraffinic hydrocarbons in the sample, and values from 15 to 20 indicate a predominance either of naphthenes or of

232

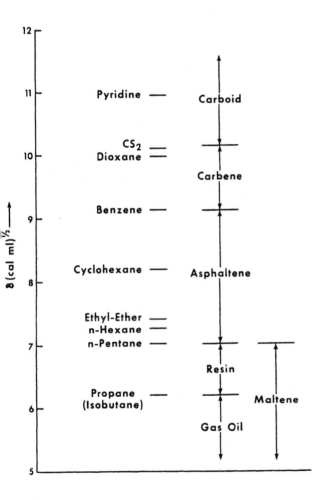

Figure 7.15: Solubility parameter ranges for solvents.

mixtures of paraffins/naphthenes/aromatics; an Index value above 50 indicates a predominance of aromatics in the fraction.

3.11 Miscellaneous Methods

Stability and incompatibility/compatibility can be estimated in terms of (1) compatibility spot tests (ASTM D 4740); (2) modified spot tests and xylene equivalents; (3) thermal stability test data (ASTM D 1661); (4) existent and potential sludge formation (hot filtration test); (5) asphaltene peptization; and (6) rate of viscosity increase of stored residual fuel oil samples at various temperatures with and without exposure to air.

In addition to the test for the stability and compatibility of residual fuels (ASTM D 4740), the other most frequently used stability test used is the hot filtration test. Hot filtration test results of up to 0.2% w/w are considered to be satisfactory; results above 0.4% w/w indicate a poor stability, but differing values might be required, depending on the intended use of the product.

One test, or property that is somewhat abstract in its application but that is becoming more popular is the **solubility parameter**. The solubility parameter allows estimations to be made of the ability of liquids to become miscible on the basis of miscibility of model compound types where the solubility parameter can be measured or calculated.

Although the solubility parameter is often difficult to define when complex mixtures are involved, there has been some progress. For example, petroleum fractions have been assigned a solubility parameter similar to that of the solvent used in the separation (Figure 7.15) although there is also the concept (Speight, 1992) that the solubility parameter of petroleum fractions may be defined somewhat differently (Figure 7.16). Whichever method is the best estimate may be immaterial as long as the data are used to the most appropriate benefit and allow some measure of predictability.

On the other hand, the solubility parameter (SP) of coal liquids has been defined by the formula (Snape and Bartle, 1984):

$$SP = 0.75 \log_{10}(M_n/200 + 0.1(\%OH_{acidic})) + 1.5C_{int}/C$$

which takes into account contributions from average molecular mass (M_n), acidic hydroxyl functions (OH_{acidic}) as a measure of polarity, and the degree of condensation of the aromatic systems (C_{int}/C) which represents the propensity for π-π bonding.

A similar arrangement can be made for calculation of the solubility parameter of shale oil but should include recognition of the nitrogen functions in the oil.

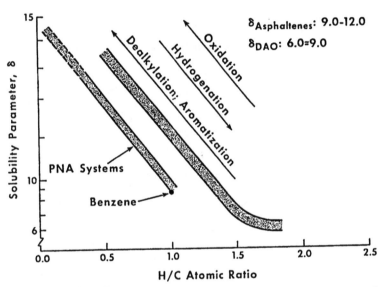

Figure 7.16: Estimation of the solubility parameter for petroleum fractions.

3.12 Bottle Tests

Bottle tests constitute the predominant test method, and the test conditions have varied in volume, type of glass or metal, use of vented and unvented containers, and type of bottle closure. Other procedures have involved stirred reactor vessels under either air pressure or oxygen pressure, and small volumes of fuel employing a coverslip for solid deposition. All of these procedures are gravimetric in nature.

A new gravimetric procedure for determining instability of

235

middle distillate fuels is termed the low pressure reactor method. In principle, it involves forcing pure oxygen, rather than compressed air, into the fuel solution at pressures up to 100 psi (794 kPa) followed by subsequent stress under conditions of accelerated storage. This method is a gravimetric determination of total insoluble materials formed.

The method is claimed to be rapid, precise, and predictive for up to three years of storage under ambient conditions. It was used to rank both additive free fuels and fuels with antioxidants added over a wide range of storage stabilities. The oxygen pressure is such that it gives a deposit of sufficient mass to make analytical determination relatively straightforward, and because of sediment mass, replicate samples offer a precise check on fuel quality.

However, there are several accelerated fuel stability tests that can be represented as a time-temperature matrix (Coordinating Research Council, 1979; Goetzinger et al., 1983). A graphical representation shows that the majority of the stability tests depicted fall close to the solid line, which represents a doubling of test time for each 10°C (18°F) change in temperature. The line extrapolates to approximately one year of storage under ambient conditions. Temperatures at 100°C (212°F) or higher present special chemical problems.

Hydroperoxide species are stable at temperatures below 100°C (212°F) but undergo both homolytic and heterolytic auto-initiated reactions at temperatures approaching this regime. Therefore, there is the inference that the stability tests might be a measure of the reactive intermediates present in a particular fuel rather than a measure of the chemical reactions occurring during storage and leading to fuel instability.

The many test matrices reported have used various fuels both neat and containing model compounds as added dopants. Likewise, simulated fuels have been studied both neat and with added dopants. Other studies have used various components isolated from fuels as dopants in otherwise stable fuels. Petroleum, coal, and shale-derived fuel liquids have been studied under these conditions, both as a source of dopants and as the fuel of interest.

The heteroatom content of the deposits formed in stability studies varied with the source or type of dopant and fuel liquid source. This result was anticipated. The color changes of both the fuel and the deposits formed are more difficult to interpret.

236

4.0 Fractional Composition

It is worthy of note that fractionation of petroleum, liquid fuels, and other products may also give some indication of stability. There are many schemes by which petroleum and related

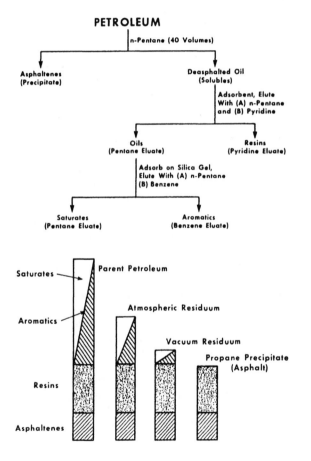

Figure 7.17: Distribution of fractions in crude oil products.

materials might be fractionated (Speight, 1991). It is not the intent to give details of these methods, since the details are available elsewhere (Speight, 1991). However, a brief overview is necessary, since fractional composition can play a role in stability and incompatibility phenomena.

Petroleum (as the example of liquid fuels and other products) can be fractionated into four broad fractions by a variety of techniques, although the most common procedure involves precipitation of the asphaltene fraction and the use of adsorbents to fractionate the deasphalted oil (Figure 7.17). The fractions are named for convenience, and the assumption that fractionation occurs by specific compound type is not quite true.

Fractionation procedures allow a before-and-after inspection of any feedstock or product and can give an indication of the means by which refining or use alters the feedstock (Speight, 1981). In addition, fractionation also allows studies to be made of the interrelations between the various fractions. For example, the most interesting phenomenon (in the present context) to evolve from the fractionation studies is the relationship between the asphaltenes and the resins.

In petroleum, the asphaltenes and resins have strong interactions to the extent that the asphaltenes are immiscible (insoluble/incompatible) with the remaining constituents in the absence of the resins (Koots and Speight, 1975). There appear to be points of structural similarity (for a crude oil) between the asphaltene and the resin constituents thereby setting the stage for a complex relationship that is more than is generally appreciated but confirming the hypothesis that petroleum is a continuum of chemical species, including the asphaltenes (Chapter 13).

This sets the stage for the incompatibility of the asphaltenes in any operation in which the asphaltene or resin constituents are physically or chemically altered. Disturbance of the asphaltene-resin relationships can be the stimulation by which the asphaltenes form a separate "insoluble" phase, leading to such phenomena as coke formation (in thermal processes) or asphalt instability, for example.

5.0 References

ASTM. 1993. American Society for Testing and Materials, Philadelphia, Pennsylvania.

Berkowitz, N. 1979. An Introduction to Coal Technology. Academic Press Inc., New York.

Butler, R.D. 1979. In Chromatography in Petroleum Analysis. K.H. Altgelt and T.H. Gouw (editors). Marcel Dekker Inc., New York.

Coordinating Research Council. 1979. CRC Literature Survey on Thermal Oxidation Stability of Jet Fuel. Report No. 509. Coordinating Research Council Inc., Atlanta, Georgia.

DeBruine, W., and Ellison, R.J. 1973. J. Pet. Inst. 59: 146.

Goetzinger, J.W., Thompson, C.J., and Brinkman, D.W. 1983. A Review of Storage Stability Characteristics of Hydrocarbon Fuels, U.S. Department of Energy, Report No. DOE/BETC/IC-83-3.

Green, L.E. 1976. Hydrocarbon Processing. 55(5): 205.

Hessley, R.K., Reasoner, J.W., and Riley, J.T. 1986. Coal Science. John Wiley and Sons Inc., New York.

Hickerson, J.F. 1975. In Special Publication No. STP 577. American Society for Testing and Materials, Philadelphia, Pennsylvania P. 71.

Koots, J. A., and Speight, J. G. 1975. Fuel. 54: 179.

Long, R.B., and Speight, J.G. 1989. Rev. Inst. Fr. du Petrole. 44: 205.

MacAllister, D.J., and DeRuiter, R.A. 1985. Paper SPE 14335. 60th Annual Technical Conference. Society of Petroleum Engineers. Las Vegas. September 22-25.

Nelson, W.L. 1964. Oil Gas J. 62(6): 124.

Nelson, W.L. 1974a. Oil Gas J. 72(2): 70.

Nelson, W.L. 1974b. Oil Gas J. 72(6): 72.

Nelson, W.L. 1976. Oil Gas J. 74(21): 60.

Nelson, W.L. 1978. Oil Gas J. 76(41): 71.

Roberts, I. 1989. Preprints, Am. Chem. Soc. Div. Pet. Chem. 34(2): 251.

Romanowski, L.J., and Thomas, K.P. 1985. Report No. DOE/FE/60177-2326. U.S. Department of Energy, Washington, D.C.

Schwartz, H.E., Brownlee, R.G., Boduszynski, M.M., and Su, F. 1987. Anal. Chem. 59: 1393.

Snape, C.E., and Bartle, K.D. 1984. Fuel. 63: 883.

Speight, J. G. 1981. The Desulfurization of Heavy Oils and

Residua. Marcel Dekker Inc., New York.

Speight, J.G. 1989. Preprints, Am. Chem. Soc. Div. Pet. Chem. 34(2): 321.

Speight, J.G. 1991. The Chemistry and Technology of Petroleum. Second Edition. Marcel Dekker Inc., New York.

Speight, J.G. 1992. Proceedings, 4th International Conference on the Stability and Handling of Liquid Fuels. US Department of Energy, Washington, DC. Volume 1. P. 169.

Speight, J.G. 1994. The Chemistry and Technology of Coal. 2nd Edition. Marcel Dekker Inc., New York.

Stuckey, C.L. 1978. J. Chromatogr. Sci. 16:482.

Thomas, K.P., Barbour, R.V., Branthaver, J.F., and Dorrence, S.M. 1983. Fuel. 62: 438.

Vercier, P., and Mouton, M. 1979. Oil Gas J. 77(38): 121.

Wallace, T.J. 1969. Advances in Petroleum Chemistry and Refining. J.J. McKetta Jr. (editor). Volume IX: Chapter 8, P. 353.

CHAPTER 8: INCOMPATIBILITY IN DISTILLATE FUELS.
I. GASOLINE AND DIESEL FUEL.

1.0 Introduction

The incompatibility of gasoline components composing the final gasoline blend is a precursor to formation of several degradation products or to other undesirable changes in the properties of the fuel.

The incompatibility of gasoline components is often the reason for a decrease in resistance to oxidation processes taking place in the gasoline blend; the resulting lower values of induction period indicate formation of free radicals ROO, which are precursors to gum formation. This situation occurs when threshold concentrations or naturally present or artificially added oxidation inhibitors (antioxidants) are lower than required.

Individually, components may be satisfactorily stable and in compliance with specifications, but in blendings may exhibit poor stability properties, making them unfit for use. As an example, assume the first component is an unstable gasoline containing a relatively high concentration of antioxidants and the second component is a stable gasoline into which only 30 ppm of a phenolic antioxidant was added. The resulting blend is affected predominantly by the unstable components, but its resulting antioxidant concentration is lower than required. Either component alone complies with specifications, but their blend does not, since its antioxidant concentration is below the effective concentration necessary for prevention of formation of excessive free peroxy radicals.

Depression of octane number values in gasoline blends occurs because of the varying octane blending values of the gasoline components or because of their differing response to octane number boosters. However, the opposite may happen: if two gasolines have the same octane number, but one contains the maximum allowed lead content and the second is lead free, then the blend of two such components usually exhibits a higher octane number than expected. This is because of the higher susceptibility of gasoline to the lower lead portions as compared with further portions of added lead: the first 0.1 g/L raises the octane number more than the second portion of 0.1 g/L (Por, 1992).

It could be argued that the differences between the expected and actual octane numbers are small, attributable to the allowed discrepancies, but experience has shown that the differences are valid and significant.

Not much attention has been paid to incompatibilities of some gasoline types when limited amounts of alcohols, especially methanol, have been incorporated. When methanol is added to paraffinic gasoline, no incompatibility problems have been observed. There are, of course, some contributions of the methanol component to the properties of gasoline containing it, but these have nothing to do with incompatibility phenomena. However, when methanol is added to olefin-rich gasoline, gum and peroxide contents are increased in comparison with each of the components alone.

Incompatibility in the middle distillate liquid fuel cut that is loosely define as diesel proves to be important to both commercial and military consumers as well as to the producer. Consumer uses, ranging from transportation to emergency auxiliary

Table 8.1   General ASTM specifications for the various types of diesel fuels.

Specification	Military[1]	No.1-D[2]	No.2-D[3]	No.1[4]	No.2[5]
°API Gravity	40	34.4	40.1	35	30
Total sulfur %	0.5	0.5	0.5	0.5	0.5
Boiling point °C	357	288	338	288	338
Flash point °C	60	38	52	38	38
Pour point °C	-6	-18	-6	-18	-6
Hydrogen (wt%)	12.5	---	---	---	---
Cetane Number	43	40	40	---	---
Acid number	0.3	0.3	0.3	---	---

1. MIL-F-16884J (1993) is also NATO F-76
2. High speed, high load engines
3. Low speed, high load engines
4. Special purpose burners
5. General purpose heating fuel oil

power supplies, require a stable, efficient fuel. This need poses both chemical and physical limitations on what refiners can produce economically.

Table 8.2: ASTM methods for determining fuel properties (see also Table 8.1).

Specification	ASTM Method
°API Gravity	D 1298
Total Sulfur %	D 129
Boiling Point °C	D 86
Flash point °C	D 93
Pour point °C	D 97
Hydrogen (wt%)	D 3701
Cetane number	D 613, D 976
Acid number	D 974

Middle distillate fuels that are made from the refining process based on straight-run distillation show very few incompatibility problems. However, at present and in the future, problems for refiners will increase as the quality of the available crude decreases worldwide. This decrease coupled with the inevitable future use of fuel liquids from both coal and shale sources will exacerbate the present problems for the producers. Currently, producers try to maximize the gasoline fraction; thus, diesel fuel is increasingly becoming a blend of cracked distillates such as light cycle oils. Care must be used in selecting these blending stocks so as to minimize incompatibility problems.

2.0 Specifications

Diesel is a liquid product distilling over the range 150-400°C (300-750°F). The carbon number ranges on average from about C_{13} to about C_{21}.

The chemical composition of a typical diesel fuel and how it applies to the individual specifications, °API gravity, distillation range, freezing point, and flash point are directly attributable to both carbon number and compound classes present in the finished fuel (Tables 8.1 and 8.2).

The color test (ASTM D-1500) is a color comparison test. Consumers buying in bulk generally prefer a lighter colored product. As a fuel degrades, it generally gets darker in color. This color change may or may not be related to degradation. Middle distillate fuels have a large proportion of the organo-nitrogen present as pyridines. Under acidic conditions, pyridines form charge transfer complexes that are highly colored. Pyridines have themselves never been implicated in incompatibility processes.

Table 8.3: Functional groups listed in order of incompatibility reactions in diesel fuels.

R-C=C- +	R-N- + Organo	R-S- + Organo-	R-O-O- Organo-
Hydrocarbons	Nitrogen	Sulfur	Oxygen
alkenes	indoles	sulfonic acids	hydroperoxides
indenes	carbazoles	thiols	dissolved oxygen
naphthalenes	pyrroles	disulfides	carboxylic acids
monocs	quinolines	sulfoxides	peroxides
polyciclics	pyridines	sulfides	aldehydes

3.0 Incompatibility Processes

Fuel incompatibility can be linked to the presence of several different deleterious heteroatomic compound classes (Chapter 6). The incompatibility observed in diesel fuel is dependent on the blending stocks employed in its production. At present, it is much more probable that any diesel fuel will be a blend containing light cycle oils and straight-run distillate.

The catalytically produced light cycle oils contain the unstable heteroatomic species that are responsible for the observed deterioration in diesel fuel. The solution is to use a straight-run distillate product, but during a deficiency in supply, an option is to use chemical additives that overcome the incompatibilities of the variant chemical composition of the light cycle oil.

INCOMPATIBILITY IN DISTILLATE FUELS
I. GASOLINE AND DIESEL FUEL

The issues that arise with the diesel fraction are much more severe than those observed in the jet fuel fraction because of the increased distillation range and the specifications allowed for the diesel fraction.

Consequently, there is a high range of interactive chemical reactions for the large number of chemical functional groups present in the diesel fuel matrix. There are many potentially different types of organic nitrogen, sulfur, and oxygen species present in any one diesel fuel (Table 8.3).

In general, the reaction sequence for sediment formation can be envisaged as being dependent upon the most reactive of the various heteroatomic species that are present in diesel fuel (Pedley et al., 1987). The worst case scenario would consist of a high olefin fuel with both high indole concentration and a catalytic trace of sulfonic acid species. This reaction matrix would lead to rapid degradation.

It is also easy to imagine the situation in which a blended fuel could pass the required accelerated test method (ASTM D-2274) at the refinery but form large amounts of insoluble material after an induction period.

Thus, just as there is no "one" diesel fuel, neither is there "one" mechanism of degradation. However, the mechanism and the functional groups involved will give a general rather than specific mode of incompatibility (Cooney et al., 1985; Hiley and Pedley, 1987; Mushrush et al., 1990, 1991; Taylor and Wallace, 1968). The key reaction in all incompatibility processes is the generation of the hydroperoxide species from dissolved oxygen. Once the hydroperoxide concentration starts to increase, macromolecular incompatibility precursors form in the diesel fuel. Acid or base catalyzed condensation reactions then rapidly increase both the polarity, incorporation of heteroatoms, and the molecular weight.

Three separate but related mechanisms can be used to explain what is occurring chemically at the different stages that a fuel goes through.

When various stocks are blended at the refinery, incompatibility can be explained by the onset of acid or base catalyzed condensation reactions of the various organo-nitrogen compounds in the individual blending stock itself. These are usually very rapid reactions with practically no observed induction

245

time period (Hardy and Wechter, 1990).

When the fuel is transferred to a storage tank or some other holding tank, incompatibility can occur with the free radical hydroperoxide-induced polymerization of active olefins. This is a relatively slow reaction because the observed increase in hydroperoxide concentration is dependent on the dissolved oxygen content (Mayo and Lan, 1987; Taylor, 1976; Mushrush et al., 1994).

The third incompatibility mechanism involves degradation when the fuel is stored for prolonged periods, as might occur during stockpiling of fuel for military use (Brinkman et al., 1980; Brinkman and Bowden, 1982; Cooney et al., 1985; Goetzinger et al., 1983; Hazlett and Hall, 1985; Stavinoha and Westbrook, 1980). This incompatibility process involves (1) the buildup of hydroperoxide moieties after the gum reactions; (2) a free radical reaction with the various organo-sulfur compounds present that can be oxidized to sulfonic acids; and (3) reactions such as condensations between organo-sulfur and -nitrogen compounds and esterification reactions.

4.0 Interactive Chemical Reactions

4.1 Oxygen and Hydroperoxide Effects

The induction time period observed for storage instability is in reality, the time required for hydroperoxide concentration to increase to the point at which the free radical induced hydro-peroxide decomposition reactions become significant (Oswald, 1961; Mushrush et al.,, 1994; Walling, 1957). These reactions are temperature dependent. If the temperature increases, both the rate of formation and, simultaneously, the rate of decomposition of hydroperoxide species increase dramatically.

Reactive olefins can either be present or generated during the thermal refining processes (Hiley and Pedley, 1987; Hartough, 1964; Wallace, 1964). Pyrolysis model studies show that even at elevated temperatures (450°C/840°F, and higher), styrene is produced from alkyl substituted benzenes (Hazlett, 1980). In a mixture of hydrocarbons, styrene was found to be resistant to further thermal reactions, thus assuring a substantial yield of active olefins in a pyrolysis stream. Other active olefins that could be present would include bicyclics such as indenes and tetralins, alkenes, and cyclic alkenes.

246

Table 8.4   Insolubles for diesel fuel with added DMP and various co-dopants at the 80°C/14 day stress matrix.

Co-Dopant*	Total Insolubles (mg/100mL Fuel)		
	Fuel D-11	Low DMP	High DMP
Hexanoic Acid - HA			
Fuel Blanks	0.3	47.4	171.4
Low HA	0.7	76.7(81.7)	160.7(252)
High HA	0.3	94.5(96.7)	259.3(282)
Acetic Acid - HOAc			
Fuel Blanks	0.0	47.4	171.4
Low HOAc	0.3	57.4(72.7)	193.8(193)
High HOAc	0.3	78.2(86.7)	145.5(264)
p-Toluene Sulfonic Acid p-TsOH			
Fuel Blanks	0.4	47.4	171.4
p-TsOH	0.4	51.0(54.7)	145.6(164.1)
Dodecylbenzene Sulfonic Acid DBSA			
Fuel Blanks	0.1	47.4	171.4
Low DBSA	0.6	373.5(382.2)	640.8(604.4)
High DBSA	0.6	1.5(6.2)	1323.6(1338.8)
N,N-Dimethylaniline DMA			
Fuel Blanks	0.1	47.4	171.4
Low DMA	0.3	56.7(55.7)	149.1(170.5)
High DMA	0.4	56.6(56.2)	124.3(167.8)
tri-n-Butylamine TBA			
Fuel Blanks	0.1	47.4	171.4
Low TBA	0.0	52.8(55.1)	145.8(164.3)
High TBA	0.1	56.7(61.3)	114.3(182.3)
4-Dimethylaminopyridine 4-DMAP			
Fuel Blanks	0.1	47.4	171.4
4-DMAP	0.0	62.9(71.5)	167.4(202.7)
Nicotinic Acid NA			
Fuel Blanks	0.1	47.4	171.4
NA	0.0	53.0(54.5)	155.4(165.0)
3-Pyridinesulfonic Acid 3-PSA			
Fuel Blanks	0.4	47.4	171.4
3-PSA	0.3	52.8(52.4)	161.7(168.7)

*Concentrations: High   Low
          DMP   =   450ppm N(w/v)   = 135 ppm N

4.2 Acid/Base Catalyzed Incompatibility

A study of the effect of acids and bases on liquid fuels is complicated by the lack of analytical methodology to select out the molecules that are actually responsible for the incompatibility reactions. This is an area in which model studies can make significant impact.

Table 8.5: Elemental analysis for sediments derived from DMP in D-11 with added co-dopants at the 80°C/14 day stress matrix.

Co-Dopant*	Percent by Weight				Empirical Formula
	N	C	H	O	
HOAC	11.37	60.35	5.09	23.19	$C_{6.2}H_{6.2}NO_{1.8}$
HA	10.99	61.67	5.50	21.84	$C_{5.6}H_{7.0}NO_{1.7}$
DA	10.82	64.25	5.88	19.05	$C_{6.8}H_{7.4}NO_{1.5}$
p-TsOH	11.16	59.55	5.09	24.20	$C_{6.2}H_{6.3}NO_{1.9}$
NA	11.12	59.60	5.06	24.22	$C_{6.3}H_{6.3}NO_{1.9}$
3-PSA	11.05	59.39	5.15	24.41	$C_{6.3}H_{6.5}NO_{1.9}$
TBA	11.17	59.61	5.41	23.81	$C_{6.2}H_{6.7}NO_{1.9}$
DMA	10.86	59.20	5.18	24.76	$C_{6.4}H_{6.6}NO_{2.0}$
4-DMAP	12.35	59.41	5.51	22.73	$C_{5.6}H_{6.2}NO_{1.6}$

*Concentrations

DMP = 450 ppm N (w/v)

HOAc, HA, DA, DMA, = $3.21 \times 10^{-2} M$
4-DMAP, and TBA

p-TsOH, NA, 3-PSA = saturated solution

Carboxylic acids increase the amount of total insolubles formed by dimethylpyrrole (DMP) (Taylor and Frankenfeld, 1978; Beal et al., 1987; Cooney et al., 1985; Mushrush et al., 1986; Por, 1992). The high acid concentration increased the amount of insolubles more than the low concentration. Acetic acid was somewhat less effective than the larger carboxylic acid, namely, hexanoic acid. Neither of these two acids stimulated formation of

Table 8.6: Insolubles for DFM (D-11) with added 3-MI and various co-dopants at the 80°C/14 day stress matrix.

Co-Dopant*	Total Insolubles (mg/100 mL Fuel)		
	Fuel D-11	Low 3-MI	High 3-MI
Acetic Acid - HOAc			
Fuel Blanks	0.0	----	0.9
Low HOAc	0.5	5.9(8.7)	15.2(32.4)
High HOAc	2.3	8.3(12.3)	6.2(47.4)
Hexanoic Acid - HA			
Fuel Blanks	0.4	----	0.9
Low HA	0.3	4.8(8.0)	25.0(39.9)
High HA	0.4	5.8(16.9)	41.0(53.7)
p-Toluenesulfonic Acid - p-TsOH			
Fuel Blanks	0.2	----	0.9
p-TsOH	0.2	0.3(0.5)	0.4(0.7)
Dodecylbenzenesulfonic Acid - DBSA			
Fuel Blanks	0.1	----	0.9
Low DBSA	1.0	0.9(0.1)	1.3(1.1)
High DBSA	0.7	2.0(1.1)	69.7(2.2)
tri-n-Butylamine - TBA			
Fuel Blanks	0.4	0.0	0.9
Low TBA	0.0	0.7(0.2)	0.1(0.3)
High TBA	0.7	0.1(0.2)	0.2(0.4)
N,N-Dimethylaniline - DMA			
Fuel Blanks	0.3	----	0.9
Low DMA	0.4	0.6(0.5)	6.9(10.3)
High DMA	0.4	0.3(0.5)	3.2(2.1)
4-Dimethylaminopyridine - 4-DMAP			
Fuel Blanks	0.0	----	0.9
4-DMAP	0.4	0.6(0.5)	0.9(0.9)
Nicotinic Acid - NA			
Fuel Blanks	0.1	----	0.9
NA	0.1	0.2(0.1)	4.9(28.2)
3-Pyridinesulfonic Acid - 3-PSA			
Fuel Blanks	0.1	----	0.9
3-PSA	0.2	0.1(0.0)	5.6(6.1)

*Concentrations: High   Low
        DMP   = 450 ppm (w/v) = 135 ppmN (w/v)

249

Table 8.7: Elemental analysis data for sediments derived from 3-MI in diesel fuel with added co-dopants at the 80°C/14 day stress matrix

Co-Dopant*	Percent by Weight				Empirical Formula
	N	C	H	O	
NONE	5.36	72.63	5.88	16.13	$C_{15.8}H_{15.2}NO_{2.8}$
HOAc	5.57	72.72	5.92	15.77	$C_{15.2}H_{14.8}NO_{2.5}$
HA	5.65	72.16	5.78	16.41	$C_{14.9}H_{14.2}NO_{2.4}$
DA	5.81	72.55	5.76	15.88	$C_{14.6}H_{13.8}NO_{2.4}$

* Concentration = all at 3.21 x 10^{-2} M

insolubles when DMP was absent (Table 8.4).

A readily soluble sulfonic acid, dodecylbenzene sulfonic acid, interacted very strongly with DMP in diesel fuel. Analysis of the sediment, combined with the fact that the yields of sediment achieved in several instances are in excess of the amount that could be attributed only to DMP participation, indicates that the acid (or possibly, molecules derived from the base fuel) is condensing directly with the pyrrole.

The incorporation of molecules of dodecylbenzene sulfonic acid into the sediment is implied by the facts that (1) such a large amount of sediment (up to 1300 mg/100 mL) was generated that the DMP alone is unable to account for the mass, (2) a number of complex absorption peaks appear in the S = O/S - O region of the infrared, and (3) the sediment is soluble in deuterochloroform and the `H NMR spectrum (90 MHz) indicates that a dodecylbenzene moiety is a major component of the sediment.

Thus, the dodecylbenzene sulfonic acid can serve as a reactant in addition to or instead of serving as an acidic catalyst. The incorporation of sulfur into DMP-derived sediments has been observed when thiophenol was used as a co-dopant. It has been considered that "in situ" oxidation of thiols to sulfonic acids may be a viable pathway to interactive behavior.

The DMP-derived sediments produced in all nine other interactive experiments proved to be remarkably similar to the insoluble material generated by DMP without a co-dopant (Tables 8.5 and 8.6). Thus, carboxylic acids are effective catalysts but are not reactants for the DMP sedimentation process.

In a similar manner, the interaction of "3-methylindole" with the 10 acid/base co-dopants was examined in diesel fuel.

With carboxylic acids, a synergism was noted insofar as the highest levels of insoluble material corresponded to higher concentrations of the indole.

Analysis of the 3-methylindole sediments was limited by the small amounts of material isolated in many of the experiments (Table 8.7). Elemental analyses could only be conducted on the sediments from the carboxylic acid interactions; all contained ca. 5.7% N, 72.5% C, and 5.8% H.

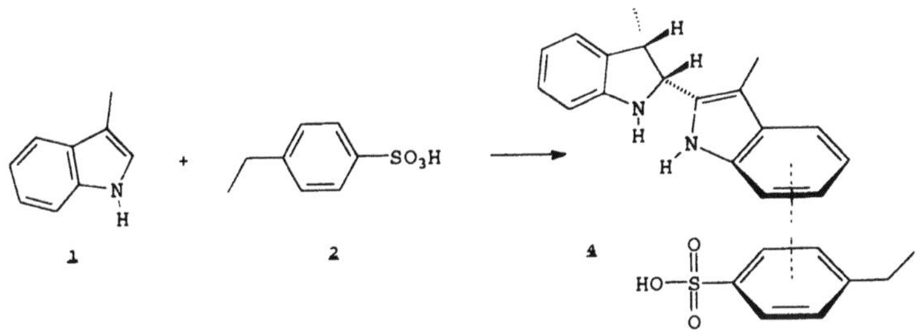

Figure 8.1: Possible Mechanism Number 1

The structure and reaction mechanism of 3-methylindole products is thus of considerable interest. The reaction is not a simple polymerization of the indole but involves the sulfonic acid as both an acid catalyst and a reactant. A recent result (Fig 8.1) shows the structure of the product and the mechanism of reaction for 3-methylindole and para-ethylbenzene sulfonic acid (Mushrush and Hardy, 1994). The product and the mechanism are depicted in the following reaction schemes.

5.0 Consequences of Incompatibility

The consequences of the reactions that can occur in fuels from manufacture to comsumer use can have serious consequenses for combustion machinery. Diesel engines are precision machines. The usual setup is to employ screens and 256 filters before the fuel injector. If low-quality fuel is used, screen and filter plugging will cause further fuel system deterioration. Partially plugged screens and filters cause changes in fuel flow. At a lower flow, the temperature increases, and nozzle plugging by fuel coking becomes a severe problem. Soluble gum constituents are also responsible for injector coking.

possibly as a π complex

INCOMPATIBILITY IN DISTILLATE FUELS
I. GASOLINE AND DIESEL FUEL

One of the incompatibility phenomena associated with diesel fuel is their behavior at low temperatures (i.e., cold properties) Cold properties of diesel fuel are defined either by their fluidity or by their filterability when cooled below certain temperatures. The corresponding tests are the pour point (ASTM D-97), the cloud point (ASTM D-2500), and the cold filter plugging point (IP 309). The first two are estimates of the fluidity, the third of the filterability. There are additional testing procedures relating to cold properties of diesel fuel.

The cold properties of diesel fuel are directly influenced by the concentration, size, and form of wax crystals that are formed when the fuel is cooled. Higher concentrations of large-size wax crystals are detrimental to the free flow of the fuel, i.e., they increase its pour point. Wax crystals of a needlelike form, even when not too large, interfere with the filterability of the fuel. Small, ball-like crystals may not affect the fluidity of the filterability, but they cause the fuel to become cloudy.

Blending two diesel fuels, both having satisfactory cold properties, may yield, under certain circumstances, a blend in which either the fluidity or the filterability is less satisfactory. This is especially true when one of the components contains pour point depressants or wax crystal modifiers, the former improving the fluidity, the latter the filterability. A possible mutual effect on the size and, especially, on the form of the wax crystals may be the reason for inferior cold properties of the blend, even when the individual components have no such problem.

Corrosivity to copper (ASTM D-130) and poor stability properties as defined by the existent gum (ASTM D-381) accelerated oxidation (ASTM D-2274), and other stability tests, can be affected by incompatibilities of diesel fuel components: corrosivity may be affected by a reduction, in the blend, of some naturally occurring corrosion inhibitors in one of the components, the corrosivity of which is prevented by such an inhibitor, but when diluted by blending with another diesel fuel component, the resulting concentration of the inhibitor in the blend may no longer be sufficient. A similar situation may arise with respect to stability properties, antioxidants taking the place of corrosion inhibitors. However, it should be noted that incompatibility regarding corrosivities and stabilities of diesel fuel components is rather rare.

A relatively high asphaltene content in one of the diesel fuel components may be a precursor to its deposition when blended with another component of a paraffinic nature. The occurrence of this type of incompatibility is rarely observed, because asphaltene concentrations in diesel fuel are generally insignificant.

INCOMPATIBILITY IN DISTILLATE FUELS
I. GASOLINE AND DIESEL FUEL

6.0 References

American Society for Testing and Materials. 1993. Philadelphia, Pennsylvania.

Beal, E.J., Cooney, J.V., Hazlett, R.N., Morris, R.E., Mushrush, G.W., Beaver, B.D. 1987. Mechanism of Syncrude/Synfuel Degradation. Report No. DOE/BC/10525-16. U.S. Department of Energy, Washington, DC.

Brinkman, D.W., Bowden, J.N., Giles, H.H. 1980. Crude Oil and Finished Fuel Storage Stability: An Annotated Review. NTIS No. DOE/BETC/RI-79/13. National Technical Information Service, Springfield, Virginia.

Brinkman, D.W., Bowden, J.N. 1982. Fuel. 61: 1141.

Cooney, J.V., Beal, E.J., Hazlett, R.N. 1984. Liquid Fuels Techn. 2: 395.

Cooney, J.V., Beal, E.J., Hazlett, R.N. 1985. Ind. Eng. Chem. Prod. Res. Dev. 24: 294.

Goetzinger, J.W., Thompson, C.J., and Brinkman, D.W. 1983. A Review of Storage Stability Characteristics of Hydrocarbon Fuels. Report No. DOE/BWTC/IC-83-3. U.S. Department of Energy, Washington, DC.

Hardy, D.R., and Wechter, M.A. 1990. Energy Fuels. 4: 270.

Hartough, H.D. 1964. Advances in Petroleum Chemistry and Refining, Wiley Interscience, New York. Pp. 419-480.

Hazlett, R.N. 1980. Frontiers of Free Radical Chemitry, .W.A. Pryor (editor). Academic Press, New York.

Hazlett, R.N., Hall J.M. 1985. Chemistry of Engine Combustion Deposits. L.B. Ebert, (editor). Plenum Press, New York.

Hiley, R.W., Pedley, J.F. 1987. Fuel 67: 1124.

Mayo, F.R., and Lan, B.Y. 1987. Ind. Eng. Chem. Prod. Res. 26: 215.

Mushrush, G.W., Beal, E.J., Pellenbarg, R.E., Hazlett, R.N., Eaton, H.R., Hardy, D.R. 1994. Energy Fuels. 8: 851.

Mushrush, G.W., Beal, E.J., Hazlett, R.N., and Hardy, D.R. 1990. Energy Fuels. 5: 258.

Mushrush, G.W., Cooney, J.V., Beal, E.J., Hazlett, R.N. 1986 Fuel Sci. Technol. Int. 4: 103.

Mushrush, G.W., Hardy, D. R. 1994. Fuel Div. Am. Chem. Soc. 39: 904.

Mushrush, G.W., Hazlett, R.N., Pellenbarg, R.E., and Hardy, D.R. 1991. Energy Fuels 5: 258.

Mushrush, G.W., Watkins, J.M., Hazlett, R.N., and Hardy, D.R. 1987. Pet. Div. Am. Chem. Soc. 32: 841.

Oswald, A.A., Noel, N.F. 1961. J. Chem. Eng. Data 6: 294.

Pedley, J.F., Hiley, R.W., and Hancock, R.A. 1987. Fuel. 66: 1645.

Por, N. 1992. Stability Properties of Petroleum Products. Israel Institute of Petroleum and Energy, Tel Aviv, Israel.

Stavinoha, L.L., and Westbrook, S.R. 1980. Accelerated Stability Test Techniques for Diesel Fuels. Report No. DOE/BC/10043-12. U.S. Department of Energy, Washington, DC.

CHAPTER 9: INCOMPATIBILITY IN DISTILLATE FUELS.
II. JET FUEL

1.0 Introduction
Most jet fuel is produced from the petroleum fraction that is generally described as the **middle distillate fraction**. However, as the quality of crude oil continues to decline and the amount of available middle distillate decreases or is changed in composition from that of three decades ago, the issue of using alternate

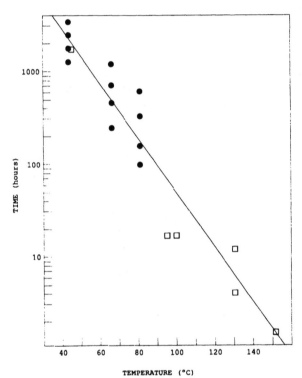

Figure 9.1: Time temperature matrix for several stability studies

sources arises. It is possible that the analogous fraction from an alternate source might be sufficiently different in composition (Table 9.1) to raise the issue of efficiency of use as well as the issues of stability and incompatibility.

Many reports using time-temperature matrices for several fuel stability tests have appeared in the literature (see, for example, Goetzinger et al., 1983; Coordinating Research Council, 1979). A

Table 9.1. Jet fuel characteristics from various sources

Source	Availability	Characteristics Positive and Negative
Petroleum	U.S. - moderate	high quality fuels
	World - moderate	low upgrading costs
		high n-alkane content
		excellent combustion quality
		transport to demand areas
Shale	U.S. - high	high quality fuels
	World - high	high n-alkane content
		excellent combustion quality
		moderate upgrading costs
		transport to demand areas
		new large capital investment
Coal	U.S. - high	low quality fuels
	World - high	low n-alkane content
		poor combustion quality
		extensive upgrading costs
		new large capital investment
Tar/Oil Sands	U.S. - low	low quality fuels
	World - low	moderate n-alkane content
		best used for a fuel blend
		poor combustion quality
		extensive upgrading costs
		large capital investment

Table 9.2.   Commercial and military specifications for jet fuels.

Fuel Types	Specification
Jet-A	ASTM D1655
JP-4	MIL-T-5624
JP-5	MIL-T-5624
JP-8	MIL-T-83133
JP-10	MIL-P-87107

graphical representation of this summary (Figure 9.1) shows that the majority of the stability tests depicted fall close to the solid line which represents a doubling of test time for each 10°C change in temperature.   The line extrapolates to approximately one year of storage under ambient conditions.   Temperatures at 100°C or higher present special chemical conditions.

Hydroperoxides are stable at temperatures of 100°C or less (Chapter 6) but undergo auto-initiated reactions, both homolytic and heterolytic, at temperatures approaching this regime.   The question becomes whether the stability tests are a measure of true storage incompatibility or a measure of reactive intermediates present in a particular fuel.   This point   is not presently resolved.

The many test matrices reported have used various fuels both neat and containing model compounds as added dopants (Goetzinger et al., 1983; Frankenfeld and Taylor, 1983).   Other studies have used various components isolated from fuels as dopants in otherwise stable fuels (Cooney et al., 1985)   Petroleum, coal, and shale-derived fuel liquids have been studied under these conditions, both as a source of dopants and as the fuel of interest (Pedley et al., 1987, 1988).

2.0 Specifications

There are a variety of specifications (Table 9.2) for jet fuel because of its use as a commercial and a military fuel.

For each jet fuel type, chemical composition, API gravity, distillation range, freezing point, and flash point are directly attributable to both carbon number and compound classes present in the finished fuel.

259

As might be anticipated, incompatibility can be ascribed to the presence of specific heteroatomic compounds. The most deleterious organic oxygen compounds present are the hydroperoxides; for organic sulfur compounds it is the sulfonic acids and/or their precursors, mercaptans and disulfides; and for organic nitrogen compounds, it is the indole or pyrrole type structures. The specifications (Tables 9.3 and 9.4) for total nitrogen and sulfur are, thus, only slight incompatibility indicators.

The move from the JP-4 jet fuel to JP-8 jet fuel is necessary but problems may be anticipated because of the higher molecular

Table 9.3   General ASTM specifications for the various types of jet fuels.

Specification	Jet A	JP-4	JP-5	JP-8	JP-10
°API Gravity	37-51	45-57	36-48	37-51	18.5-20
Total sulfur %	0.3	0.3	0.4	0.3	----
Boiling point °C	300	320	320	330	----
Flash point °C	38	---	60	38	55
Freezing point °C	-40	-72	-40	-58	-79
Aromatics (wt%)	20	25	25	25	---
Acid number	0.1	0.01	0.015	0.01	---

weight and different composition of the JP-8 jet fuel. The higher molecular weight is best illustrated by the boiling ranges of the two fuels, which for JP-4 is 60-225°C (135-440°F) and for JP-8 is 165-265°C (330-510°F). The higher boiling point of the JP-8 jet fuel will mean that higher temperatures are necessary for vaporization. Thus, there is the real potential for thermal changes to the fuel, leading to instability/incompatibility of the components.

3.0 Incompatibility Processes
Incompatibility reactions in jet fuel involve chemical components that are more complex than the gasoline components but

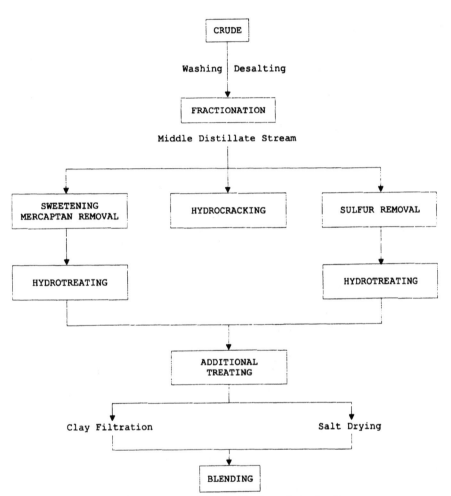

Figure 9.2: General process for refining jet fuel.

somewhat less complex than the components of the higher-molecular-weight fuels and other products, such as asphalt.

Table 9.4: ASTM methods for property specifications (see also Table 9.3).

Specification	ASTM Method
°API Gravity	D1298
Total sulfur %	D1266, D1552, D2622
Boiling Point °C	D86
Flash Point °C	D56, D3243
Freezing Point °C	D2386
Aromatics (wt%)	D1319
Acid Number	D974, D3242

To a large extent, the refining process (Figure 9.2) controls the quantity of deleterious compounds present in the finished fuel. It would, however, be prohibitively expensive to remove all (100%) of the deleterious organic nitrogen, oxygen, and sulfur moieties in the refining process. The major reaction in any hydrotreating process is for desulfurization, and if total nitrogen reduction is also a necessity, then the quantity of hydrogen required in hydrotreating is proportional to the amount and chemical type of organic nitrogen and sulfur present in the crude source.

Nitrogen removal requires both more severe operating conditions and a greater quantity of hydrogen. However, it is the removal of these two functional groups that dramatically improves fuel compatibility. The quantity of hydrogen consumed per barrel is related to the crude source (Table 9.5), and hydrogen

Table 9.5: Jet fuel source and hydrotreating data.

Process Pressure	Hydrogen Consumption	Purpose
Low	2-3 ft³/BBL	Mercaptan removal
Low	40-60	Sulfur reduction
High	200-300 Shale	Nitrogen Removal
High	200-400 Petroleum	Aromatic Saturation
High	400-800 Coal	Aromatic Saturation

consumption is indicative of the quality of the source crude and its subsequent refining costs (Speight, 1981).

In general, the order of heteroatom removal depends on bond energy, and since the carbon-nitrogen bond is much stronger than the carbon or carbon-sulfur bonds, this is reflected in both the quantity of hydrogen and process severity required to lower organic nitrogen content.

4.0 Interactive Chemical Reactions

Thiophenol, a reactive thiol that readily forms the thiyl radical in an oxidizing environment, was employed in a study (Watkins et al., 1989) to illustrate the oxidizing reactions of the peroxide species. Neither thiophenol nor its oxidized product, phenyl disulfide, is masked by the other components present in a fuel matrix. Triphenol was, thus, the sulfur compound of choice.

The results from a series of control samples (Table 9.6) show that peroxidation occurred in a cyclic pattern. Similar oscillations had been noted previously (Watkins et al., 1989). In fact, the control sample (Jet A blending stocks) formed peroxides at a greater rate than the Shale JP-5 or the nonblended (Jet A) samples. It might be suggested that the low-molecular-weight blending stocks lacked the natural inhibitors present in the true fuel samples.

It is important to note that the fuels doped with sulfur (in the form of thiophenol) did not undergo peroxidation as rapidly as the fuel-only samples. Thiophenol eliminated all peroxide species present and, when the added sulfur concentration was halved, peroxide formation began earlier in the fuels that did exhibit extensive peroxidation.

This finding demonstrated a relationship between added sulfur and peroxide formation (Shertzer, 1978; Watkins et al., 1989). However, it might be more accurate to term this process "peroxide inhibition." Under the same conditions, hydrofined blending stock showed evidence of peroxide formation throughout the duration of the tests. For sulfur-doped fuels, all of the samples showed dramatic peroxide reduction except the jet fuel derived from shale oil which required a longer period for the peroxide concentration to reach zero.

However, the shale oil derived sample contained a larger

Table 9.6: Jet fuel peroxidation at 65°C with added thiophenol.

WEEK	SHALE-II JP-5		JET A		HYDROCRACKED JET A BLENDING STOCK		HYDROFINED JET A BLENDING STOCK	
	CONTROL	DOPED	CONT.	DOPED	CONT.	DOPED	CONT.	DOPED
0.10% Added Sulfur								
0	0.25	0.25	0.00	0.00	0.16	0.16	0.00	0.00
1	0.24	0.00	0.19	0.00	0.57	0.00	0.18	0.00
2	0.31	0.00	0.44	0.00	1.16	0.00	0.49	0.00
3	0.37	0.00	0.19	0.00	1.73	0.00	1.10	0.00
4	0.51	1.29	0.40	0.51	5.38	0.26	4.08	0.00
5	0.48	0.97	0.26	0.40	8.47	0.25	10.82	0.00
0.05% Added Sulfur								
0	0.69	0.00	0.12	0.00	0.24	0.00	0.10	0.00
1	0.70	0.00	0.18	0.00	1.58	0.00	0.60	0.00
2	0.73	0.00	0.16	0.00	6.01	0.00	2.09	0.00
3	0.94	0.45	0.28	0.54	37.66	0.61	12.41	0.00
4	1.11	0.68	0.26	0.22	62.05	0.51	25.27	0.00
5	1.56	0.88	0.29	0.81	59.92	0.25	56.67	0.00
0.03% Sulfur								
0	0.60	0.60	0.01	0.01	0.27	0.27	0.06	0.06
1	0.62	0.14	0.04	0.00	0.96	0.00	0.32	0.00
2	0.72	0.00	0.05	0.00	1.88	0.00	0.78	0.00
3	0.84	0.00	0.00	3.98	4.20	0.00	2.46	0.00
4	0.90	0.01	0.09	0.01	10.04	0.01	5.89	0.00
5	0.93	0.00	0.05	0.50	22.43	0.00	13.26	0.00
6	0.99	0.00	0.00	0.70	49.03	0.00	28.54	0.00
7	1.14	0.03	0.13	1.38	77.88	0.62	60.62	0.00
8	1.29	0.15	0.07	0.88	97.40	0.36	77.44	0.00
0.01% Added Sulfur								
0	0.51	0.51	0.00	0.00	0.25	0.25	0.06	0.06
1	0.60	0.20	0.10	1.04	1.02	0.73	0.50	0.00
2	0.67	0.17	0.13	0.72	2.37	0.11	1.32	0.00
3	0.81	0.00	0.20	0.89	4.66	0.00	3.63	0.00
4	0.85	0.00	0.17	0.55	13.22	0.00	13.79	0.00
5	0.99	0.00	0.21	0.59	29.20	0.22	29.10	0.00
6	1.08	0.23	0.17	0.42	55.33	0.14	48.57	0.00
7	1.12	0.33	0.19	0.72	89.57	0.19	78.27	0.00
8	1.51	0.46	0.24	0.65	131.13	0.28	124.98	0.00

initial concentration of peroxides. Even when the added sulfur was lowered by a factor of 10, peroxide control was noted, but a longer period was required to decrease the peroxide levels to zero. The added sulfur was not enough to control the large amount of peroxide formation in the hydrocracked blending stock sample.

The two Jet A samples, both control and doped, formed peroxides at a greater rate than the shale oil derived fuel or the nonblended fuels. This indicates that hydrotreating tends to remove the natural inhibitors, such as sulfur species, resulting in much higher observed peroxide levels for these fuels.

Model studies have shown that disulfides are the product

Table 9.7: Sulfur concentration vs Time for added sulfur dopant from thiophenol.

0.01% Added Sulfur				0.03% Added Sulfur			
Shale JP-5		Jet A		Shale JP-5		Jet A	
week	mg S	week	mg S	week	mg S	week	mg S
0	0.2	1	0.20		00.6	0	0.60
1	0.03	1	0.00	1	0.13	1	0.17
2	0.02	2	0.00	2	0.05	2	0.03
3	0.00	3	0.00	3	0.06	3	0.01
4	0.00	4	0.00	4	0.05	4	0.01
5	0.00	5	0.00	5	0.05	5	0.01
6	0.00	6	0.00	6	0.03	6	0.01
7	0.00	7	0.00	7	0.03	7	0.01
8	0.00	8	0.00	8	0.00	8	0.01

0.10% Added Sulfur				0.05% Added Sulfur			
week	mg of S/2mL	week	mg of S/2mL	week	mg of S/2mL	week	mg of S/2mL
0	2.20	0	2.08	0	1.00	0	1.00
1	1.16	1	0.98	1	0.76	1	0.25
2	0.51	2	0.81	2	0.23	2	0.33
3	0.43	3	0.63	3	0.15	3	0.16
4	0.26	4	0.47	4	0.02	4	0.15
5	0.24	5	0.41	5	0.00	5	0.00

derived from thiyl radicals under mild oxidizing conditions and at

265

moderate temperatures (Oswald and Wallace, 1966). With high concentrations of added thiophenol dopant in fuels, phenyl disulfide is an observed product, and a decrease in thiophenol sulfur concentration with time is observed (Tables 9.6 and 9.7).

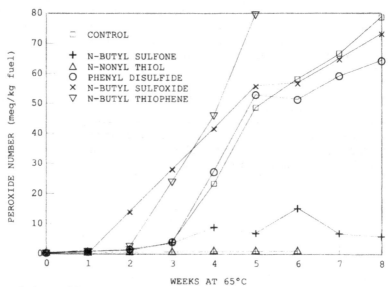

Figure 9.3: Sulfur compound reactions with naturally-occurring peroxides.

There are means by which the inconsistencies observed with deposit formation and sulfur compound interactions can be explained (Pedley and Hiley, 1986; Pedley et al., 1987, 1988). Thus, the stability of an unstable middle distillate fuel can be improved by treatment with sodium hydroxide. The subsequent addition of a sulfonic acid restores the instability of the base washed fuel. A linear relationship between deposit formation and sulfonic acid concentration confirms that the formation of a strong acid is a limiting factor in storage instability. Further, the deposits generated by the sulfonic acid treatment were found to be identical from those formed to the fuel itself on storage.

It has also been reported (Pedley et al., 1988) that the chemical structures observed in the sediment precursors consisted of indole systems linked to a phenalene ring system. In the particular diesel fuel studied, phenalene was detected. Oxidation products, i.e., various phenalenones, were observed to increase in concentration as the fuels were aged under ambient storage conditions. This appears to be a major step forward in the chemistry of sediment formation. However, more study is needed to extend this observation to other middle distillates.

These results and those from other studies have indicated that some sulfur compounds might be inhibitors for controlling hydroperoxide formation in jet fuels (Fettke, 1983; Coordinating Research Council, 1979).

To this end, samples of various aviation turbine fuels that undergo extensive peroxidation were doped with different classes of organic sulfur compounds at 0.03% added sulfur and analyzed for hydroperoxide formation during a 65°C (150°F) eight-week stress period (Cooney et al., 1985). The organic sulfur compounds employed as dopants in this study included n-nonyl thiol, phenyl disulfide, n-butyl sulfoxide, and 2-butyl thiophene (Figure 9.3). This alkyl thiol dopant was dramatic in its control of hydroperoxide formation and eliminated measurable hydroperoxides for all reaction time periods.

The results for the sulfoxide show a general lowering of peroxides after an initial period in which the peroxide level rises slightly, agreeing with the general chemical behavior of alkyl sulfoxides in model systems. In model studies a sulfoxide once formed is stable for short reaction times but shows a gradual decrease in yield as the corresponding sulfone shows a similar increase in yield (Mushrush et al., 1987). The results for the alkyl sulfone also mimic model system behavior. Alkyl sulfones are quite resistant to oxidation even at elevated temperatures.

The alkyl substituted thiophenes are some of the most oxidation resistant sulfur compounds, and a negative synergism may operate to increase peroxidation above control levels (Figure 9.3). Likewise, this finding is supported by model system studies (Nishioka, 1988).

The effect of temperature on the course of organic sulfur compound oxidation can be conveniently shown from results obtained from the jet fuel thermal oxidation stability test (JFTOT). From a study of sulfur dopants in hydrotreated kerosene (Figure 9.4), the results at first glance appear to be reasonable (Savaya et al., 1988). It was concluded that the addition of most sulfur compounds increased deposit formation but there are some discrepancies. Problems arise from the observation that varying the alkyl chain of a sulfide produces measurable differences in deposit formation, which is presumably related to the relative energies of bond dissociation.

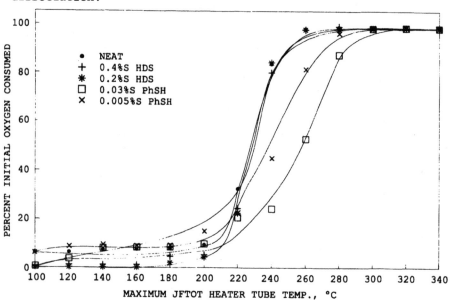

Figure 9.4: JP-5 peroxidation at 65°C in the presence of sulfur dopants

Further, the thiophene results do not agree with other literature reports (Mohammed et al., 1986). A major shortcoming in the JFTOT is the general lack of monitoring in both oxygen concentration and/or hydroperoxide depletion. Most reported jet fuel tests for thermal oxidation stability tests concern themselves with either tube deposits or filter plugging. There are

Table 9.8: Hydroperoxide concentrations measured in dodecane
containing organosulfur compounds as a function of
JFTOT maximum heater tube temperature.

Maximum Tube		Hexyl Disulfide		Thiophenol	
Temp. °C	Neat	0.2% S	0.4% S	0.005% S	0.03% S
21-120	nd*	nd	nd	nd	nd
200	0.11	0.01	nd	nd	nd
240	2.15	1.09	1.50	0.13	nd
280	2.25	1.03	1.52	1.02	nd
320	1.60	0.36	nd	0.69	nd

*not detected, less than 0.01 meq/kg.

some reports of model heteroatom studies and/or simultaneous
monitoring of active oxygen species (Hazlett et al., 1977; Hazlett
and Hall, 1985).

This is an unfortunate situation, since deposits produced by
thermal stress are the consequence of free radical autoxidation
reactions. Hydroperoxides are the most active oxygen species
present in the fuel matrix and can reach quite high values even
before stressing.

If a fuel is sparged with air for 15 min, the peroxide value
can be 0.7 meq/kg, but with pure oxygen for a similar time period
the ROOH value is raised to 2.8 meq/kg (Hazlett et al., 1977;
Morris and Mushrush, 1989). The regimen of limited oxygen
availability and short reaction time is similar to the environment
in many parts of an aircraft fuel system.

The oxidation of disulfides and thiols in a model fuel,
dodecane, in a modified JFTOT (Table 9.8) (Morris and Mushrush,
1989) showed that the addition of 0.4% (w/w) sulfur from n-hexyl
disulfide suppressed peroxidation by about 46% of the level
attained in the neat fuel. Addition of 0.2% sulfur from the same
compound resulted in suppression to approximately 68% of the neat
dodecane. Comparison of oxygen profiles (Figure 9.4) reveals that
direct oxidation by dissolved oxygen did not play a role in
peroxide reduction. Consequently, the disulfide was reacting with
free radicals in solution to reduce the formation of
hydroperoxides. The results are consistent with an oxidation

269

mechanism that involved coordination of oxygen to the sulfur centers, as discussed earlier (Field, 1977).

Thiophenol was observed to be a much more reactive sulfur dopant (Table 9.8) (Morris and Mushrush, 1989). Based on model studies reported at lower temperatures, it was not surprising to find thiophenol such an effective radical trap. The data (Figure 9.3) confirm that the free radical chain reactions were being terminated quickly before significant autoxidation could ensue.

5.0 Consequences of Incompatibility

Contrary to expectations, incompatibility phenomena are often observed in jet fuels. The properties most affected when incompatible kerosene fractions are blended are corrosivity to copper and silver strips, gum content, thermal stability, water separating capacity, and color. Incompatibility of jet fuel components is caused mainly by the various treatment methods yielding jet fuels of different characteristics.

Hydrotreating removes almost all the sulfur and nitrogen compounds from the jet fuel, converting them to hydrogen sulfide and ammonia, which are ultimately removed from the system. Hydrotreating also deprives the jet fuel of the small concentrations of inherent antioxidants and lubricity improving substances; the latter also act as corrosion inhibitors.

Hydrotreated products therefore have the following characteristics: (1) they are susceptible to free radical and peroxy radical formation; (2) they tend to form gum and other degradation products; (3) they exhibit color instability; (4) they exhibit sometimes lower water separating capacities; (5) they often have lower thermal stabilities; and (6) they may be corrosive to copper and/or silver due to small amounts of residual sulfide or free sulfur.

In order to improve the properties of jet fuel, additives are used extensively. Incompatibility problems may therefore be encountered when additives are diluted by blending operations or when incompatible additives incorporated in jet fuel components affect each other.

Acid treatment, and other chemical treatments, converts mercaptans to disulfides, but does not extract sulfur compounds present originally in the untreated product. Such acid-treated or

270

chemically treated jet fuels are rather stable and exhibit satisfactory thermal stabilities, but water separating capacities are sometimes low due to some residual detergent property associated with the possible presence of some naphthenates and sulfonate.

Regenerative soda treatment sweetens or partially extracts mercaptans from the product, depending on which stage oxidation takes place. The extraction occurs when oxidation is prevented in the mercaptans-to-mercaptides conversion stage and is allowed to take place at the later mercaptides-to-disulfides conversion stage; disulfides, insoluble in the aqueous phase, are separated and removed from the system.

The description of the various treatment processes and their effect on the characteristics of the treated jet fuels explains why blending kerosene may result in blends exhibiting properties not complying with routine quality requirements, even when the separate components are in compliance.

A good example in this respect is the corrosivity to copper and/or silver strips. A chemically treated jet fuel may contain small, but sufficient quantities of corrosion inhibitors and be, therefore, noncorrosive even if it contains some other copper/silver strip corrosive compounds. When added to an almost sulfur-free jet fuel, which in itself is also noncorrosive, a mixture is obtained in which the particular sulfur compounds acting as corrosion inhibitors are diluted below a threshold value, making the blend corrosive to the copper and/or the silver strip, depending on the concentration and the types of the sulfur compounds present in the system.

Stability properties (such as gum content, thermal stability, peroxide formation rate, and resistance to oxidation) are affected by blending incompatible components, mainly due to dilution of antioxidants necessary for keeping the less stable component within the required stability limits; this is so because antioxidant concentrations are additive, while stability properties are not. Less stable components have a larger effect on the blend than the more stable components.

The water separation characteristics may be affected similarly. One component may exhibit a satisfactory water separating capacity but be susceptible to forming a system in which water separates with difficulty and the other component may have a high water separating capacity but contain additives that interfere with that capacity. A blend of these two components may, thus, exhibit poor water separation characteristics due to the influence of additives.

6.0 References

Cooney, J.V., Beal, E.J., and Hazlett, R.N. 1985. Ind. Eng. Chem. Prod. Res. Dev. 24: 294.

Coordinating Research Council. 1979. Determination of the Hydroperoxide Potential of Jet Fuels, CRC Report No. 559. CRC Inc., Atlanta, Georgia.

Fettke, J.M. 1983. Organic Peroxide Growth in Hydrotreated Jet Fuel and its Effects on Elastomers. Report GE TM83AEB1154. General Electric, Lynn, Massachusetts.

Field, L. 1977. Organic Chemistry or Sulfur Compounds. S. Oae (editor). IFI/Plenum, New York. PP. 347-352.

Frankenfeld, J.W., and Taylor, W.F. 1983. Ind. Eng. Chem. Prod. Res. Dev. 22: 622.

Goetzinger, J.W., Thompson, C.J., and Brinkman, D.W. 1983. A Review of Storage Stability Characteristics of Hydrocarbon Fuels, Report No. DOE/BETC/IC-83/3. U.S. Department of Energy, Washington, DC.

Hazlett, R.N., Hall, J.M., and Matson, M. 1977. Ind. Eng. Chem. Prod. Res. Dev. 16: 171.

Hazlett, R.N., and Hall, J.M. 1985. Chemistry of Engine Combustion Deposits. L.B. Ebert (editor). Plenum Press, New York. P. 245.

Mohammed, A.A., Savaya, Z.F., and Abbas, K.J. 1986. Pet. Res. 5: 21-25.

Morris, R.E., Mushrush, G.W. 1989. Preprints. Div. Fuel Am. Chem. Soc. 34: 538.

Nishioka, M. 1988. Energy Fuels 2: 214.

Oswald A.A., and Wallace, T.J. 1966. Anionic Oxidation of Thiols and Co-oxidation of Thiols with Olefins. in Organic Sulfur Compounds. Volume 2. N. Kharasch and C.Y. Meyers (editors). Pergamon Press, New York. PP. 205-232.

Pedley, J.F., and Hiley, R.W. 1986. Formation of Insolubles During Storage of Naval Fuels. Proceedings of the 2nd International Conference on Long Term Storage Stabilities of Liquid Fuels, San Antonio, Texas.

Pedley, J.F., Hiley, R.W., and Hancock, R.A. 1987. Fuel. 66: 16451.

Pedley, J.F., Hiley, R.W., and Hancock, R.A. 1988. Fuel 67: 1124.

Savaya, Z.F., Mohammed, A.A., and Abbas, K. 1988. Fuel 67: 673.

Shertzer, R.H. 1978. Aircraft Systems Fleet Support/Organic

Peroxides in JP-5 Investigation, Final Report NAPC-LR-78-20. Naval Air Propulsion Center, Trenton, New Jersey.

Speight, J.G. 1981. The Desulfurization of Heavy Oils and Residua. Marcel Dekker Inc. New York.

Watkins, J.M., Mushrush, G.W., Hazlett, R.N., and Beal, E.J. 1989. Energy Fuels 3: 231-236.

CHAPTER 10: INCOMPATIBILITY IN NON-DISTILLATE PRODUCTS.
I. LUBRICATING OIL AND WAX

1.0 Lubricating Oil

After developing kerosene, the early refiners focused on paraffin wax for the manufacture of candles and lubricating oils, which were, initially, by-products of wax manufacture. The preferred lubricants of the 1860s were lard oil, sperm oil, and tallow. However, as the evolution to heavier machinery increased, the demand for lubricating oil increased, and after the 1890s, petroleum displaced animal and vegetable oils as the sources of lubricating oil (Speight, 1991).

The term **lubricating oil** (Chapter 5) represents a group of oils serving many purposes. Each of the lubricating oil types has its own special use and, therefore, its own specification requirements (Berkley, 1973; Mills, 1973).

For example, there are lubrication oils used in steam turbines, internal combustion engines, marine diesel engines, industrial and railway engines, and gear transmissions; hydraulic oils for power transmission; cutting oils for lubrication and cooling during metal working; transformer oil used as insulators in electrical equipment; and spindle oils used in textile manufacturing. Each oil is defined by a standard that prescribes the stability of the oil under service conditions (ASTM, 1993).

Thus, there is a difference between stability considerations regarding petroleum products such as gasoline, jet fuel, and diesel fuel and those of lubricating oil. In the case of liquid fuels the main stability considerations relate more to long-term storage and less to breakdown during use.

On the other hand, lubricating oil is exposed to extremely severe conditions during use, such as high temperatures in humid air, contact with metals of varying compositions, as well as exposure to contaminants. In addition, thin films of lubricating oil are exposed at the piston ring and cylinder surfaces to degradation induced by high temperatures, high pressures, and high shear stresses.

The stability testing methods and the stability estimates for lubricating oil are, therefore, adapted for measuring resistance to changes in the composition and properties during use.

Table 10.1: Viscosities of various lubricating oils.

	Kinematic viscosity, cs @ 100°C (212°F)		Dynamic viscosity, cp
	minimum	maximum	maximum
SAE 10W	4.1		3500 @ -20°C (-4°F)
SAE 15W	5.6		3500 @ -15°C (5°F)
SAE 20W	5.6		4500 @ -10°C (14°F)
SAE 25W	9.3		6000 @ -5°C
SAE 20	5.6	9.3	
SAE 30	9.3	12.5	
SAE 40	12.5	16.3	
SAE 50	16.3	21.9	

1.1   Production

Lubricating oil is derived mostly from vacuum distillates obtained from paraffinic crude oil (Chapters 2 and 5) and involves (1) preparation of the suitable fractions by vacuum distillation processes; (2) solvent refining with suitable solvents like furfural; (3) dewaxing with the help of solvents like methyl-ethyl-ketone (MEK), or methyl-iso-butyl-ketone (MIBK), or any other suitable solvent at low temperatures; (4) refining of the product, usually by mild catalytic hydrogenation; and (5) blending for obtaining the required viscosity product and addition of additives, usually additive packages (Speight, 1991; Sequeira, 1992).

Higher-viscosity lubricating oil (**bright stock**) is obtained from residual fractions by solvent extraction processes, such as the propane deasphalting process, which was originally designed to produce lubricating oil stacks in addition to asphalt.

1.2   Classification and Properties

Each of the lubricating oil types has its own quality requirements, according to the particular use of the specific oil.

Lubricating oil for use in internal combustion engines has a special importance from the point of view of use under constantly extreme conditions. Transformer oil must possess a high dielectric constant, and contaminants must be absent; transformer oils must exhibit a degree of stability. Cutting oil must possess a high emulsion stability and they must not contain components that might be corrosive. Hydraulic oil must be resistant to fouling and must have a good water separating capacity; absence of degradation products and other contaminants is important. Spindle oil must be well refined and exhibit sufficiently low viscosities.

The main functions of lubricating oil for internal combustion engines are to reduce friction and to avoid "seizure" of the moving metallic surfaces by producing a thin film of the lubricant acting as a barrier between moving surfaces. Additional functions are cooling, sealing, exclusion of contaminants, ensuring cleanliness and protection against wear.

Engine lubrication oil is classified by viscosity (ASTM D-445) (Table 10.1) based on the Society of Automotive Engineers (SAE) J 300 specification. The single-grade oils (SAE 20, SAE 30, etc.) correspond to a single class and must be selected according to engine manufacturers' specifications and operating and climatic conditions. Multigrade lubrication oils (SAE 10W, SAE 20W, etc.) cover several SAE grades; for example, the SAE 10W/30 grade possesses at -20°C (-4°F) the viscosity of a 10W oil and at 100°C (212°F) the viscosity of an SAE 30 oil. The required viscosity index is obtained by adding a suitable additive to improve the viscosity index.

The **viscosity index** (ASTM D-2270) is one of the key properties of engine lubricating oil and indicates the rate of change of viscosity of lubricating oil at different temperatures. The lower the rate of change, the better is the lubrication oil quality. The index value of 100 indicates the smallest viscosity change with temperature; the value of 0 indicates the largest viscosity change. The viscosity index (VI) can be calculated as follows:

$$VI = (L - U)/(L - H) \times 100$$

where U is the viscosity of the oil at 40°C (104°F); L is the viscosity of the oil of VI = 0, its viscosity at 100°C (212°F) being equal to the viscosity of the oil; H is the viscosity of the oil of VI = 100, its viscosity at 100°C (212°F) being equal to the viscosity of the oil.

Naphthenic oils usually have higher viscosity-temperature coefficients than do paraffinic oils at equal viscosities and temperatures. The viscosity index requirements in most of the lubricating oil specifications are ca. 95. The index is one method of estimating lubricating oil stability; for example, a lubricating oil performing well under high-temperature conditions would have a high viscosity index.

The absence of inorganic acidity is mandatory; total acid numbers (ASTM D-974) equivalent to 0.05 mg potassium hydroxide per gram are allowed in most of the specifications. Pour points (ASTM D-97) of -6°C to -9°C (23°F down to 16°F) are usually required.

Viscosity (ASTM D-445, ASTM D-4741), **density** (ASTM D-1298), **carbon residue** (ASTM D-524), **color** (ASTM D-1500), **flash point** (ASTM D-92), **ash content** (ASTM D-482), and **sulfur content** (ASTM D-92) requirements are included in many of the lubricating oil specifications, the limitations depending on the type and on the particular lubricating oil grade.

Reduction of flash points; increases in ash contents, carbon residues, and water contents; darkening of color; and decrease of viscosity indicate a degradation of the lubricating oil properties during operation.

1.3   Composition
Lubricating oil is composed mainly of hydrocarbons of an approximate molecular weight range of the order of 420 to 520 and a boiling range of 350-450°C (660-840°F), depending on the particular lubricating oil grades as classified by viscosity.

The general tendency is to incorporate in the lubricating oil as high a proportion of paraffinic hydrocarbons and as little as possible of the naphthenic components, particularly aromatic components. This is normally achieved by employing suitable refining processes, such as furfural extraction which removes most of the aromatic components.

Typical physical property analyses of lubricating oil (van Nes and van Westen, 1951) show the following proportions of hydrocarbon groups in the final product:

> carbon in paraffinic chains: 65 - 75%
> carbon in naphthenic rings: 20 - 30%
> carbon in aromatic rings: <5%

In addition to the hydrocarbons, lubricating oil also contains sulfur, nitrogen, and oxygen compounds in small concentrations. Some of these compounds are present in the product originally, and others are added to the finished product.

Total sulfur content (in the form of thiols, sulfides, and thiophenes) may be as high as 2%. Nitrogen compounds, in much smaller concentrations, are represented by quinolines and pyridines (basic nitrogen compounds that are extracted by mineral acids), pyrroles, indoles, and carbazoles. Oxygenated compounds are represented mainly by organic acids.

In spite of the low concentrations of these heteroatom constituents, they have a pronounced effect on the stability properties of the lubricating oil.

1.4   Stability

The oxidation stability of lubricating oil can be significantly improved if the lubricating oil is produced from suitable crude oils, such as paraffin-based crude oils.

However, the supply of such crude oils has diminished over the past decades, and the predominant feedstocks are now naphthenic in nature. Feedstocks containing naphthenic-aromatic components are inadequate for lubricating oil production, both in terms of the yield and the quality of the lubricating oil. However, improvements in the lubricating oil quality from such feedstocks can be achieved by use of inhibitors containing oxidation and corrosion inhibitors, antiwear additives, detergents and dispersants, viscosity index improvers, pour point depressants, and antifoam agents.

Achieving good stability properties is essential if oxidation processes are to be prevented. These processes result in thickening of the oil (increased viscosity, which influences performance) and the formation of degradation products that can cause machine damage.

Resins and asphaltenes, as defined by fractionation techniques (Chapter 1) and which occur through degradation of the oil, are precursors to sludge formation which causes engine failures due to ring and valve sticking. Oxidation of oil alone or in the presence of metals ions leads to resin and asphaltene formation and (by inference) increased viscosity; the formation of sludge or sediment is the overall outcome. High temperature oxidation results in the formation of products such as alcohols, aldehydes, and acids by oxidation of the organic components of the oil, simply:

alcohol formation:
$$R\text{-}CH_2\text{-}R' + [O] = R\cdot CHOH\text{-}R'$$

aldehyde formation:
$$R\text{-}CHOH\text{-}R' + [O] = R\text{-}HC{=}O\text{-}R' + H_2O$$

ketone formation:
$$R\text{-}CHOH\text{-}R' + [O] = R\text{-}CO\text{-}R' + H_2O$$

where R and R' are hydrocarbon moieties.

Since the breakdown of lubricating oil is often associated with oxidation, many of the testing procedures are based on oxidation under conditions involving high temperatures and the presence of oxygen. One such procedure is the test for the oxidation characteristics of the oil (ASTM D-943). Another stability test is the oxidation stability of steam turbine oil by the rotating bomb method (ASTM D-2272).

Resistance to oxidation of extreme pressure fluid lubricants, gear oils or mineral oils is often determined by a procedure (ASTM D 2893) in which the oil is exposed to 95°C (205°F) in the presence of dry air for 312 hours. The amount of precipitated material and increase in the kinematic viscosity are determined.

The determination of the **sludging tendencies** of lubricating oil during oxidation in the presence of oxygen, water, copper, and iron can also be determined by a standard method (ASTM D-4310).

In this method, an oil sample is reacted with oxygen in the presence of water and an iron-copper catalyst at 95°C (205°F) for 1000 hours. The stability of the oil is estimated from the weight of insoluble material, which is determined gravimetrically by filtration of the oxidation tube contents through 5-μm pore size filter discs. Change in the acid number also serves as an additional estimate.

A test method (ASTM D-4636) for the oxidation stability of hydraulic oil, turbine engine lubricating oil, and other highly refined lubricating oil is available, and a procedure for oxidation stability of automotive engine oil by the thin film oxygen uptake method (ASTM D-4742) is also available.

However, most of the laboratory aging tests are actually accelerated tests due to the fact that they allow the performance of the oil to be evaluated within an acceptable period of time. However, caution is advised, since these tests use higher temperatures and more severe exposures to oxygen. Thus, the use and/or the interpretation of the data may be open to question and

Table 10.2: Lubricating oil additives

Antioxidants:
Zinc dialkyldithiophosphate, Aromatic amines, alkylated phenols.
Metal deactivators:
Amines, organic complexes containing nitrogen or sulphur, disalicylidene diaminopropane, sulphides, phosphites.
Anti-foams:
Silicon polymers, organic copolymers.
Corrosion and rust inhibitors:
Zinc dialkyldithiophosphate, metal phenolates, amines and fatty acids, basic metal sulphonates.
Detergents:
Sulphonates, phosphates and phenolates.
Dispersants:
Alkylsuccinimides and polymeric alkylthiophosphates.
Friction modifiers:
Amines and fatty acids, lard oil.
Anti-wear and extreme pressure agents:
Zinc dialkyl dithiophosphate, organic phosphates, sulphurized fats, organic sulphur and chlorine compounds, sulphides and disulphides.
Pour point depressants:
Chlorinated paraffins condensed with naphthalene, phenolic polymers, polymetacrylates, phenolic naphthalene.
Viscosity index improvers:
Polymers and copolymers of butadiene, methacrylates, alkylated styrenes.

---

may not always be representative of the conditions of use and the ensuing deterioration of the oil, although some correlations helpful to determining stability are possible (Murray et al., 1983).

In order to correlate such accelerated laboratory tests with the actual resistance of lubricating oil to the conditions of use, full-scale engine oxidation tests are necessary. Such engine tests assess not only oil thickening and conventional degradation products formation, but also the rate of coke, resin, asphaltene, and sludge formation. Rig tests are carried out by testing sequences as specified in the relevant specifications at defined oil temperatures, the test conditions depending on the particular method used. Among the parameters established in such engine tests are the thermal stability, rate of deposit formation, wear and corrosion rate, and valve clogging.

1.5   Additives

The main additives used to enhance lubricating oil performance are oxidation inhibitors, friction modifiers, metal deactivators, extreme pressure additives, foaming inhibitors, pour point depressants, corrosion and rust inhibitors, viscosity index improvers, detergent, and dispersants.

Most of these additives are used in commercial lubricating oils. The composition and concentrations differ according to manufacturers' specifications, but are representative of several basic additive compositions (Table 10.2) (Garcia-Borras, 1986).

Considering the extremely severe operational conditions to which lubricating oil is exposed and at which oxidative processes are active, oxidation inhibitors play a most important role. Sterically hindered phenols are often used, alone or together with other antioxidants or metal deactivators such as, for example, disalicylidene diaminopropane. Compatibility of the constituents and a properly synergistic action of such an antioxidant mixture must be ensured.

The inhibition mechanism is based on abstraction of hydrogen from the phenolic molecule by the chain propagating peroxy radicals to form radicals that are stabilized by resonance and interfere with the propagation stage of the chain reaction.

Metal deactivators interfere with the action of metallic ions which are active in decomposing hydroperoxides and which initiate

Table 10.3: Classification of lubricating grease

Consistency	Worked penetration
1	310 - 340
2	265 - 295
3	220 - 250
4	175 - 205
5	130 - 160

oxidation chain reactions.

When evaluating lubricating oil base stocks and refining methods, stability considerations must be kept well in mind. Base stocks should have a convenient chemical composition, and suitable refining methods should be used for producing products of chemical compositions that would resist oxidative processes. Additives can also be used for improving the properties and performance of lubricating oils and for increasing their resistance to degradation processes.

2.0   Lubricating Grease

A **grease** (Chapter 5) is a lubricant composed of a liquid lubricating oil and a thickening agent, such as a soap of calcium, aluminum, barium, lithium, sodium, and other metals (Barnes et al., 1973; Dawtrey, 1973; Speight, 1991).

Soap is made by chemically reacting a metal hydroxide with a fat or fatty acid. Fats are chemical combinations of fatty acids and glycerine:

$$3R\text{-}CO_2H \; + \quad \begin{array}{c} CH_2OH \\ CHOH \\ CH_2OH \end{array} \quad = \quad \begin{array}{c} CH_2OCOR \\ CHOCOR \\ CHOCOR \end{array}$$

$$\text{fatty acid} \qquad \text{glycerine} \qquad\qquad \text{fat}$$

and the subsequent reaction is simply represented:

$$R\text{-}CO_2H + NaOH = RCO_2^-Na^+ + H_2O$$

The most common metal oxides used for this purpose are sodium hydroxide, calcium hydroxide, lithium hydroxide, and barium hydroxide. Fillers such as graphite or metal oxide powders are sometimes added in order to impart special properties to the greases.

The instability/incompatibility of grease, like that of lubricationg oil, is often associated with deterioration during service and this is very true to reality. However, greases differ markedly in their tendency to liberate oil, and opinions differ as to whether the lubricating properties of a grease depend upon this ability. Nevertheless, the liberation of oil from the grease during storage (ASTM D-1742) or in feed lines is to be avoided.

2.1   Composition and Classification

The composition of lubricating grease can be measured by application of the appropriate test (ASTM D-128), and properties such as viscosity (ASTM D-1092) can also be measured.

Grease is classified either according to the type of or the consistency of the thickening agents (i.e., the soap), which determine the suitability of the grease to perform satisfactorily under the specified working conditions. The five commonly used consistency grades are defined by their worked penetration ranges (ASTM D-217) (Table 10.3).

Grease is used where a semisolid, high-viscosity lubricant is preferred to a liquid lubricant, when exclusion of humidity and dust is necessary, in cases where adhesion to metal surfaces is to be ensured, and in cases of high and intermittent pressure changes.

The type of a grease necessary is determined according to the particular nature of its use and the working conditions to which it is intended to be exposed. Conditions such as the velocity of the moving metal surfaces, pressure, temperature, humidity, and contaminants, whose contact with the metal surfaces has to be minimized, are given consideration prior to the choice of a suitable grease.

2.2   Properties

Many of the general properties and specification requirements of lubricating greases may be considered as stability requirements. Changes in any of the original properties, including consistency changes, if affected by environmental influences or by storage conditions, are in fact associated with instability phenomena. In

ING OIL AND WAX

Table 10.4: An illustration of the effect of the various soaps
on some of the properties of the greases:

Type of soap	Lifetime	Temperature (°F)	Resistance to water	Stability
Aluminum	short	175	fair	poor
Barium	intermediate	350	good	good
Calcium	intermediate	175	fair	fair
Lithium	long	300	good	high
Sodium	long	200	weak	good

addition, some direct stability estimates are also included in quality requirements of lubricating greases.

Specifications include definitions of the thickening agents (type of soaps), which is actually a stability requirement. The type of thickening agent is closely associated with (1) resistance of the grease to consistency changes during use, and (2) resistance to softening and to phase separation during storage and use (Table 10.4). Among some of the direct stability considerations are resistance to oxidation, age hardening, phase separation, color stability, and resistance to consistency changes.

Resistance to oxidation is established usually by the oxygen bomb method (ASTM D-942). The consistency of grease at high temperatures (ASTM D-3232) is a critical parameter, which is indicated by the **equivalent viscosity**. This is a torque measurement expressed in units of viscosity (poises).

Phase separation (a sign of instability or incompatibility) is usually estimated by tests to determine the amount of oil separated by centrifuging (ASTM D-4425) or by the amount of oil separated during storage (ASTM D-1742).

Cone penetration (ASTM D-217) is an estimate of the consistency of the grease, indicated by its hardness as such and retention of its penetration after application of strokes (usually 10,000 double strokes), and observing the penetration increase (change in the consistency indicated by the softening of the product, which is usually limited to 15% maximum).

The dropping point (ASTM D-566 or ASTM D-2265) indicates whether the thickener tends to melt or whether oil separates at extreme temperature ranges.

The viscosity of the liquid component is usually determined (ASTM D-1092) in order to supply information about the grade of the lubricating oil component used for the grease.

Among other tests employed in monitoring the quality of lubricating grease are the water content (ASTM D-95), ash content (ASTM D-128), acidity and alkalinity (ASTM D-974) and copper strip corrosion (ASTM D 4048).

3.0 Petroleum Wax

Although it is generally accepted that lubricating oil was a by-product from wax manufacture (discussed above), there is also the suggestion that wax (as it is known at present) is actually a by-product of lubricating oil manufacture (Mazee, 1973). Be that as it may, and it is not the purpose of this text to question the origins of products, both stories are true.

Wax is a product of the petroleum (Chapter 5) and other industries that has found wide use in a variety of areas, the most common being in the manufacture of candles. Although there is an expanding use for wax in the petrochemical industry, where the hydrocarbon components make excellent stocks for (to mention only two alternate products) gasoline and diesel fuel.

3.1 Classification

Wax can generally be subdivided (classified) into three types: paraffin wax, microcrystalline wax, and petrolatum, also known as petroleum jelly.

Paraffin wax is usually obtained by dewaxing refined lubricating oil stocks. Dewaxing processes usually employ a combination of chilling, filtration, and extraction by suitable solvents like MEK or MIBK (Scholten, 1992).

Soft wax is obtained at the preliminary dewaxing stages and, therefore, contains some residual quantities of oil. **Hard wax** is obtained by de-oiling the soft wax. In order to ensure color stability and absence of odor, the wax is subjected to a sulfuric acid or an active earth treatment.

286

3.2 Properties

One of the properties of wax that can cause problems during use is the potential for the wax to crystallize from solution in response to a drop in temperature. This is, essentially, the basis of the process for wax separation by crystallization.

Wax may also phase separate from the surrounding liquid medium when the aromatic content of the liquid increases beyond a critical point, the point being determined by the wax content and the other constituents of the liquid.

Thus, the presence of wax constituents in various petroleum products may have an adverse effect on the performance characteristics of the product.

In terms of lubricating oil performance, wax deposition at the "wrong time" can cause seizure of machine parts. Another example of wax deposition having an adverse effect on performance occurs when diesel fuel is used in extreme temperatures such as exist in northern climates, and the fuel deposits wax components in the lines. One method of combatting such a potential occurrence is to maintain the fuel lines above the freezing point (0°C; 32°F) in a warm enclosure or by allowing the engine to run even when the vehicle is parked.

There are many property specifications of wax that have been documented (Gottshall and McCue, 1973) and are necessary for use of wax as a commercial product. However, several of these

Table 10.5: Properties of microcrystalline waxes.

Drop melting point	ASTM D 127	70-76°C (160-170°F)
Oil content	ASTM D 721	1% wt. max.
Needle penetration (25°C)	ASTM D 1321	17 mm/10 max
Kinematic viscosity (100°C)	ASTM D 445	12-16 cs
Flash point (Clev. open cup)	ASTM D 92	225°C (435°F) max.
Liquid density at 85°C	ASTM D 1298	0.795-0.805
Inorganic acidity	ASTM D 974	nil

specifications present indications of the properties of the wax as these properties will influence the stability of the wax in service.

The congealing point (ASTM D 938) and the oil content (ASTM D-721) of wax are basic indications as to the effect of temperature on their consistency or as to the temperature at which a particular wax will pass from the solid state to the liquid state, or vice versa. The congealing point of hard wax is usually between 50 and 65°C (130-150°F), depending on the grade, while the oil content is usually between 0.5 and 1.0% w/w.

The color and color stability (ASTM D 1500) are affected by the refining treatment of the wax and should be stable at a range of 0.5 to 1 (ASTM color according to ASTM D-1500). Inorganic acidity (ASTM D-974) is limited by some specification to 0.03-0.05 mg potassium hydroxide per gram. Kinematic viscosity at 100°C (212°F) (ASTM D 92) is usually in the range of 220-245°C (430-475°F). The needle penetration at 25°C (77⁰F) (ASTM D-1321) is another indication of the consistency of wax; the desirable range is 15-18 mm/10.

Microcrystalline wax originates in waxy crude oils or residual products and is obtained by combined cold settling, cold filtration and extraction processes. This type of wax differs from other waxes in the rheological properties; some exhibit the ductility properties of plasticity, and some grades are hard and even brittle (Table 10.5).

Petrolatum (Chapter 5) is a type of microcrystalline wax and, in the highly refined state, is known as vaseline. Petrolatum contains both a liquid and a solid phase, and the physical structure is similar to a colloidal system in which the solid hydrocarbons constitute the dispersed phase and the liquid hydrocarbons are the dispersing phase.

In addition to testing procedures similar to those described above, the ultraviolet absorbance and absorptivity (ASTM D 2008) of the refined petrolatum is determined at 290 nm.

4.0 References

ASTM. 1993. Annual Book of ASTM Standards. American Society for Testing and Materials, Race Street, Philadelphia, Pennsylvania.

Berkley, J.B. 1973. In Criteria for Quality of Petroleum Products. J.P. Allinson (editor). Halsted Press, Toronto. Chapter 10.

Barnes, R.S., Goodchild, E.A., and Wyllie, D. 1973. In Criteria for Quality of Petroleum Products. J.P. Allinson (editor). Halsted Press, Toronto. Chapter 11.

Dawtrey, S. 1973. In Modern Petroleum Technology. G.D. Hobson and W. Pohl (editors). Applied Science Publishers Inc., Barking Essex, England. Chapter 21.

Garcia-Borras, T. 1986. Chemtech. December. p. 130

Gottshall, R.I., and McCue, C.F. 1973. In Criteria for Quality of Petroleum Products. J.P. Allinson (editor). Halsted Press, Toronto. Chapter 12.

Mazee, W.M. 1973. In Modern Petroleum Technology. G.D. Hobson and W. Pohl (editors). Applied Science Publishers Inc., Barking Essex, England. Chapter 20.

Mills, A.L. 1973. In Modern Petroleum Technology. G.D. Hobson and W. Pohl (editors). Applied Science Publishers Inc., Barking Essex, England. Chapter 20.

Murray, D.W., McDonald, J.M., White, A.M., and Wright, P.G. 1983. Proceedings. 11th World Petroleum Congress.

Scholten, G.G. 1992. In Petroleum Processing Handbook. J.J. McKetta (editor). Marcel Dekker Inc., New York. P. 565.

Sequeira, A. Jr. 1992. In Petroleum Processing Handbook. J.J. McKetta (editor). Marcel Dekker Inc., New York. P. 634.

Speight, J.G. 1991. The Chemistry and Technology of Petroleum. 2nd Edition. Marcel Dekker Inc., New York.

van Nes, K., and van Westen, H.A. 1951. Aspects of the Constitution of Mineral Oils. Elsevier, Amsterdam.

CHAPTER 11: INCOMPATIBILITY IN NONDISTILLATE PRODUCTS.
II. RESIDUAL FUEL OIL

1.0 Production and Classification

Fuel oils are also called heating oils and are distillate products that cover a wide range of properties. The specifications for such products are not as demanding as for other products (Hoffman, 1992).

Although fuels oils have been defined elsewhere (Chapter 5), at the risk of some repetition, for the purposes of this chapter, it is worth redefining the different fuel oils to make clear the differences between distillate fuel oil and residual fuel oil.

No. 1 fuel oil is very similar to kerosene and is used in burners where vaporization before burning is usually required and a clean flame is specified. **No. 2 fuel oil** is often called **domestic heating oil** and has properties similar to diesel fuel and heavy jet fuel; it is used in burners where complete vaporization is not required before burning. **No. 4 fuel oil** is a light industrial heating oil and is used where preheating is not required for handling or burning; there are two grades of No. 4 fuel oil, differing in safety (flash point) and flow (viscosity) properties. **No. 5 fuel oil** is a heavy industrial fuel oil that requires preheating before burning. **No. 6 fuel oil** is also a heavy fuel oil and is more commonly known as **Bunker C oil** when it is used to fuel ocean-going vessels; preheating is always required for burning this oil.

Residual fuel oil is obtained by blending residual products from various refining processes with suitable diluents, usually middle distillates, to obtain the required fuel oil grades. The residual products are obtained from atmospheric distillation, vacuum distillation, visbreaking, and catalytic cracking (Gruse and Stevens, 1960; Por, 1992; Gray, 1994).

Residual fuel oil is usually classified according to viscosity. For example, a light fuel oil is characterized by a viscosity (ASTM D-445) of 62 cs at 50°C (122°F), a medium grade fuel oil by a viscosity of 175 cs at 50°C (122°F), a heavy fuel oil by a viscosity of 325 cs at 50°C (122°F), and extra heavy fuel oil has a viscosity exceeding 325 cs at 50°C (122°F).

2.0 Properties

Fuel oil used for domestic purposes or for small heating installations would be of the lower viscosities and lower sulfur contents (Kite and Stephens, 1973; Walmsley, 1973). In large-scale industrial boilers, heavier grade fuel oil would be used, with sulfur content (ASTM D-1552 or ASTM D-129) requirements regulated according to the environmental situation of each installation and the local environmental regulations (Pope, 1973).

The flash point (ASTM D-93) is usually limited to 60°C (140°F) minimum because of safety considerations. Requirements for asphaltene content (IP 143), carbon residue value (ASTM D 189, ASTM D-524), ash (ASTM D-482), water content (ASTM D-95), and metals content are included in some specifications.

The pour point (ASTM D-97), indicating the lowest temperature at which the fuel will retain its fluidity, is limited in the various specifications according to local requirements and fuel handling facilities: the upper limit is sometimes 10°C (50°F); in warm climates it is somewhat higher.

Another important specification requirement is the heat of combustion (ASTM D-240), the specified values usually being 10,000 Cal/kg (gross) or 9400 Cal/kg (net).

Because of economical considerations, residual fuel oil has been replacing diesel fuel for marine purposes. Viscosity specifications had to be adjusted to the particular operational use, and some additional quality requirements also had to be allowed for. The main problems encountered during the use of residual fuel oils for marine purposes are the stability properties (sludge formation) and, even more so, the compatibility properties of the fuel.

Blending of fuel oils for obtaining lower viscosity values and mixing fuel oils of different chemical characteristics are sources for deposit and sludge formation in the vessel fuel systems. This incompatibility is mainly observed when high asphaltene fuel oils are blended with diluents or other fuel oils of a paraffinic nature. Stability properties and compatibilities of fuel oils, the mechanisms involved, the suggested testing methods, and the recommended remedies are described in detail below.

3.0 Stability and Incompatibility

The specification requirements of the various fuel oil blends are attained by blending the residues obtained at the bottom of atmospheric distillation towers (straight-run residues), vacuum residues, catalytic cracker residues, and thermal cracker/visbreaker residues with suitable diluents; the diluents are mostly middle or heavy distillates obtained in the various straight-run, vacuum, or cracking operations. The effect of the chemical composition of the diluents is of major importance in respect to the stability and incompatibility/compatibility of the resultant fuel oil (Kite and Stephens, 1973; Por, 1992).

Residual fuel oil stability and compatibility (ASTM D-4740) is often affected by the severity of the cracking operation to which the fuel oil has been subjected. Stability problems associated with residual fuel oils appear in the form of sludge and/or a solid deposit that settles at the bottom of a storage tank or plugs filters in fuel systems. Sludge and deposit formation is associated not only with product instability but also with incompatibility.

However, studies of the various mechanisms pertaining to the stability properties and the compatibilities of residual fuel oils indicate that estimation of the peptization state of the asphaltene micelles would well indicate the degradation formation rate in residual fuels (Por et al., 1988).

The increasing need for producing more distillates from each barrel of crude oil necessitates employment of more severe conversion processes for increasing distillate yields. Alternative energy sources, diminishing the residual fuel demands, also contribute to the need of increasing production of distillates at the expense of residues. Therefore, more severe visbreaking processes will have to be applied, adjusting the yield demand but resulting in residual fuels of a lesser stability.

Vacuum residua, when serving as fuel oil base stocks rather than as visbreaker feedstock or asphalt feedstock, have to be blended with diluents because of viscosity requirements; compatibility problems may be expected in this case, especially when relatively high asphaltene content stacks are blended with diluents of an insufficient aromatic reserve.

Deposition of sludge or sediment may also be expected in blends of two or more residual fuels, forming a combined continuous phase of too low a peptizing power. The use of fuel oil in marine engine systems, rather than marine diesel fuel or middle distillates, contributed to problems of fuel system clogging by excessive formation of sludge and deposits, due in most of the cases to poor compatibilities of the system components.

Stability and incompatibility/compatibility can be estimated in terms of (1) compatibility spot tests (ASTM D-4740), (2) modified spot tests and xylene equivalents, (3) thermal stability test data (ASTM D-1661), (4) existent and potential sludge formation (hot filtration test), (5) asphaltene peptization; and (6) rate of viscosity increase of stored residual fuel oil samples at various temperatures with and without exposure to air.

In addition to the stability and compatibility test of residual fuel oil (ASTM D-4740), the other most frequently used stability test is the existent and accelerated dry sludge content determination of residual fuels (the hot filtration test). Hot filtration test results of up to 0.2% w/w are considered to be satisfactory. Results above 0.4% w/w indicate a poor stability, but differing values might be required, depending on the intended use of the product.

The tendency of a fuel oil to form degradation products may also be estimated by the thermal stability test (ASTM D-1661). There are certain limitations to this test, mainly in respect to viscosities: if the viscosity is not too high, this thermal stability test can be used; it indicates satisfactorily the fouling tendency of the tested fuel.

The ratings indicate stable, borderline, and unstable products (codes 1, 2, 3 correspondingly), according to the appearance of the sludge and deposits on the thimble.

The tendency of free radical formation in various residues by different initiators of polymerization chain reactions has been investigated in this respect. The initiators used are heat, ultraviolet irradiation at wave lengths near the visible range (since C-H or saturated carbon-carbon bonds are not to be affected), accelerated oxidation, and catalytic action of metals. The free radical formation is then correlated with the behavior of the various types of residual fuels during storage and use.

Compatibility can be defined as the absence of suspended solids in residual fuel oil blends under the conditions of the test. The compatibility rating is an assigned numerical value of

294

a reference spot produced under the conditions of the test by allowing a drop of the fuel oil, or a drop of the fuel oil and diluent blend, to spread on chromatographic paper of specified grade. The procedure differs according to the particular method used, and the numerical values range from 1 to 5, the lower numerals indicating better compatibilities.

The compatibility of various fuels with each other or with diluents can be determined (Por, 1992). Almost any diluent is compatible with catalytically cracked fuel oils or fuels of a high aromaticity; on the other hand, only diluents of an aromatic character are compatible with straight-run residues. This is, of course, to be expected, since diluents of a paraffinic character will precipitate asphaltene from fuel oil. The asphaltene content of catalytically cracked residues (being a product of heavy vacuum gas oil) is low, and therefore, it is compatible with almost any diluent.

Straight-run residues are assumed to be one of the more stable residual products; visbroken residues are the least stable. The thermal stability of catalytically cracked residues was found to be intermediate. Regarding blends with diluents, straight-run diluents are preferable in this respect to cracked diluents.

In order to prevent or to diminish the formation of degradation products in residual fuel oils and their blends, two controversial influences have to be considered: there is better compatibility in the case of catalytically cracked residues and catalytically cracked diluents, the latter containing significant proportions of aromatic components. On the other hand, these products have a poorer stability than straight-run residues. These straight-run residues and straight run diluents have good stability, but there are problems in respect to the compatibilities of their blends.

The main reasons for these particular stability and compatibility properties can be explained by the availability (or non-availability) of a sufficient aromatic reserve in conjunction with the micellar structure of the dispersed phase, which is a part of the colloidal solution characterizing the residual fuel oil structure.

4.0 Nature of the Degradation Products
The instability or incompatibility of residual fuels is

characterized by formation of separate phases, which usually take one or more of the following forms: (1) precipitation of coke-like (carbonaceous) particle or inorganic impurities, both insoluble in aromatic solvents (ASTM D-4484); (2) separation of waxy materials; (3) precipitation of oxidation or polymerization products; (4) precipitation of asphaltenes (ASTM D-3279); and (5) increase in the viscosity of the residual fuels during storage.

The asphaltene content of residual fuel oil is assumed to affect stability and compatibility (Por, 1992;) (see also Chapter 13).

Asphaltenes are complex mixtures and the molecular structure of the constituents is not uniform (Chapter 13). Thus, asphaltenes differ in their response and susceptibility to environmental influences as well as in their solubilities in hydrocarbons. They exist in petroleum in the form of a micelle, in which the nuclear components are surrounded by resin constituents, which in turn, are surrounded by aromatic components. The dispersed and the dispersing phases are in equilibrium, which can be easily shifted in either direction. Such a shift may result in a change of the residual fuel stability or compatibility.

The asphaltene contents of visbroken residua are usually much higher than those of the paraffinic vacuum residua, while their peptization values are significantly lower. Again, the asphaltene content increases with time, while the peptization values remain approximately constant. The higher asphaltene content of visbroken residua, in comparison with those of vacuum residua, is not surprising. This is a result of the concentration of the asphaltenes in the product. The asphaltenes themselves are not destroyed by the visbreaking process, while some of the other components are converted to distillates that are separated from the asphaltene-containing residue.

The degree of peptization of the asphaltene constituents can serve as an estimate of fuel oil stability. The higher peptization values indicate a fuel oil of higher stability. In this situation the peptization capacity of the intermicellar phase is high enough to ensure a sufficient peptization of the micelles, and precipitation of asphaltenes is avoided. Blending stocks and diluents having a high aromaticity maintain the asphaltene micelles in a fully peptized state.

INCOMPATIBILITY IN NON-DISTILLATE FUELS
II. RESIDUAL FUEL OIL

The equilibrium of this system may be easily disturbed either by changes of temperature or by blending with less aromatic liquids. The degree of peptization state of the asphaltene micelles becomes impaired, and flocculation is initiated. This situation can be avoided by using blending stocks or diluents of sufficient peptizing power, i.e., which have a high aromaticity.

Components having a high wax content tend to disturb the equilibrium of the system and to support formation of degradation products.

In the case of visbroken residua, the degree of peptization of the asphaltene micelles (which may be the inherent micelles as well as micelles formed during the cracking process) may also affect the stability of the fuel.

The susceptibility to oxidation processes is also associated with fuel instability, resulting in deposit formation (see for example, high hot filtration test values of cracked residual fuels (Por, 1992)). During storage, oxidation and polymerization products may be formed that, if soluble in the fuel, will cause a viscosity increase. If these products are insoluble, they will settle out as highly viscous, semisolid or solid phases.

The chemical composition of residual fuel oil may well be assumed to be one of the main factors influencing the stability and the compatibility.

Since the viscosity increase of residual fuel oil during storage indicates a change of the original properties, its rate of change may be considered as an estimate for the degree of fuel oil instability. In the case of residual fuel oil, the increase of viscosity as a result of degradation products formation is not only a quality problem, but also an economical disadvantage, since the price of residual fuel oil is based on a reference viscosity. If the fuel oil viscosity rises, more distillate material is needed to reduce it to the reference viscosity level, and this decreases the value of the fuel oil.

The rise in viscosity of stored fuel oil is generally due to degradation products that remain in solution; these products are formed during storage due to the temperature and/or the presence of aerial oxygen. If the chemical characteristics of the fuel oil are favorable to the dissolution of the degradation products, the viscosity of the fuel oil increases, but sludge or deposit formation is often absent. Such fuel oil is often considered to be stable, since there is no obvious sign of sludge or deposits. In fact, however, the degradation products are dissolved in the fuel oil.

Additional factors affecting sludge and deposit formation in fuel oil during storage include (1) precipitation of waxy matter; (2) effect of water; and (3) the presence of extraneous solids such as silt, sand, and dust. Emphasis must also be placed on the composition, storage time, storage temperature, and extent of exposure to oxygen.

An interpretation of the colloidal behavior of residual fuel oils, the peptization state of asphaltenes or the tendency of the asphaltene micelles to remain peptized or to settle out and form sludgy deposits, seems to be a suitable approach to the stability and compatibility problem relating to residual fuel oil.

One approach is to base the instability/incompatibility of heavy residual fuels on the chemical composition as well as on the internal colloidal structure (Por, 1992) by defining a colloidal instability index (CII) as a ratio of the sum of asphaltenes and saturated oils to the sum of the peptizing resins and aromatic solvents:

CII = (Asphaltenes + Saturates)/(Aromatics + Resins)

The equilibrium of a well-peptized asphaltene system can be easily disturbed by application of heat, oxidation, ultraviolet irradiation, and addition of a paraffinic diluent. In each case, the chemical composition is altered and the aromaticity is decreased, thereby causing a disruption of the equilibrium of the colloidal system.

As a result, asphaltene particles are deprived of their enveloping layers, which previously merged continuously with the consecutive layers. The micelle system becomes noncontinuous, and the asphaltene cores are prone to agglomeration. Such a process leads to an instability of residual fuel oils, taking the form of sludge and deposits at the bottom of storage tanks and of formation of degradation products clog lines, filters, and fuel systems.

When studying degradation products from residual fuel oil, care must be taken to distinguish between the following two phenomena: (1) degradation products formed due to instability, and

(2) degradation products formed due to incompatibility.

Unstable fuels, originating mostly from cracking/visbreaking processes, tend to form free radicals, leading to polymerization chain reactions, which result ultimately in formation of degradation products. When these are soluble in the fuel oil or when the fuel oil, due to its chemical composition, is capable of dissolving them, its viscosity is increased.

When the degradation products do not dissolve in the fuel oil, either because of their nature or because the fuel oil is not able to dissolve them, they settle out as sludge or deposits. Often both these processes occur simultaneously insofar as part of the degradation products is dissolved (with an accompanying increase in fuel oil viscosity) and part remains undissolved and forms sludge and deposits.

Degradation products that formed due to the incompatibility of some of the fuel oil components are associated with a different mechanism. When two or more fuels, incompatible with each other, are blended, the degradation products are formed by a chemical reaction between two or more chemical groups present in the fuel oil components. In most cases, the products either settle as a sludge or float in the fuel oil, thus forming a heterogeneous solution.

Incompatibility in residual fuel oil is often due to a high asphaltene content of one of the components in the blend, especially when such a component is blended with another fuel of a predominantly paraffinic nature or when the high-asphaltene component is blended with a paraffinic diluent. In this case, the asphaltenes are precipitated by the paraffinic constituents (Speight, 1991).

However, not only the concentration of the asphaltenes, but also their peptizibility is of importance in this respect.

Most estimates of the potential for compatibility and/or incompatibility in residual fuels, when blended with diluents are based on determinations of formation rates of sludge and suspended solids.

In the compatibility spot test (ASTM D-4740) a drop of the preheated fuel oil sample is placed in an oven at 100°C (212°F), and after 1 hour the spot is examined for evidence of precipitation; at this stage a stability estimate is obtained. As a second step, a blend consisting of equal amounts of the fuel oil and the blending stock is handled similarly: a compatibility estimate is so obtained.

In another spot test (ASTM D-2781), compatibility ratings are based on numerical values of a reference spot obtained under the particular testing procedure in which a drop of the fuel oil and diluent blend is allowed to spread on a chromatographic paper of a specified grade. The exact procedures vary according to the particular method used. Assigned estimates from 1 to 5 indicate the homogeneity of the obtained spot, 1 being the most homogeneous spot, indicating the best compatibility, and 5 showing a very dark spot surrounded by light colored areas, indicating very poor compatibility.

Another procedure, sometimes called the xylene equivalent test, is based on determining the proportion of xylene, in a blend with iso-octane, necessary for obtaining a homogeneous spot on chromatographic paper when blended with an equal quantity of the tested fuel oil sample. The lower the proportion of xylene in the diluent, the better is the compatibility of the tested fuel.

A modified spot test for determining the compatibilities of fuel oil is based on using a diluent composed of equal proportions of n-heptane and toluene or of two types of gas oils, one a straight-run (paraffinic) and the second a catalytically cracked (olefinic-aromatic) gas oil. The spots obtained by drops of the fuel oil and diluent blend on chromatographic paper are evaluated as in the other spot tests, but this modified test is at a higher temperature (usually 63°C; 145°F) indicating asphaltene deposits are indicated.

The test for sludge formation (hot filtration test) is useful for stability as well as for compatibility estimations. In this test, final fuel oil blends yielding in this test values below 0.3% would indicate good stability properties as well as acceptable compatibilities.

Data for the peptization state of asphaltenes are useful in compatibility estimations. High peptizibility, as indicated by the lower values obtained in the procedure, points to a stable colloidal system in which micelles are homogeneously integrated. Such a system would exhibit good compatibility.

Viscosity increases during fuel oil storage at various temperatures, with and without contact with air, are a good estimate of the compatibility of such blends. Viscosity increases indicate formation of degradation products that have been dissolved in the fuel.

Since not only incompatibility, but also instability of fuels are responsible for such degradation products formation, care should be taken when interpreting the results. When samples are kept out of contact with air, the effect of instability is reduced. In this case, incompatibility would be the main cause for such a viscosity increase of the fuel oil blend during storage.

5.0 References

Gray, M.R. 1994. Upgrading Petroleum Residues and Heavy Oils. Marcel Dekker Inc., New York.

Gruse, W.A., and Stevens, D.R. 1960. Chemical Technology of Petroleum. McGraw-Hill, New York. Chapter XII.

Hoffman, H.L. 1992. In Petroleum Processing Handbook. J.J. McKetta (editor). Marcel Dekker Inc., New York. P. 2.

Kite, W.H., and Stephens, G.G. 1973. In Criteria for Quality of Petroleum Products. J.P. Allinson (editor). Halsted Press, Toronto. Chapter 9.

Pope, J.G.C. 1973. In Modern Petroleum Technology. G.D. Hobson and W. Pohl (editors). Applied Science Publishers, Barking Essex, England. Chapter 18.

Por, N. 1992. Stability Properties of Petroleum Products. Israel Institute of Petroleum and Energy, Tel Aviv, Israel.

Por, N., Brauch, R., Brodsky, N., and Diamant, R. 1988. Proceedings, 3rd International Conference on the Stability and Handling of Liquid Fuels, London.

Speight, J.G. 1991. The Chemistry and Technology of Petroleum. 2nd Edition. Marcel Dekker Inc., New York.

Walmsley, A.G. 1973. In Modern Petroleum Technology. G.D. Hobson and W. Pohl (editors). Applied Science Publishers, Barking Essex, England. Chapter 17.

CHAPTER 12: INCOMPATIBILITY IN NONDISTILLATE PRODUCTS.
III. ASPHALT

1.0 Introduction

There is a certain confusion regarding the nomenclature of the heavier feedstocks, especially asphalt and bitumen (Chapter 1), and various systems of nomenclature have been suggested (Table 12.1) (Abraham, 1945; Hanson, 1964; Speight, 1991). Pitch and tar are, more correctly, the products of the thermal decomposition of coal, but the terms are often applied to thermal asphalt.

In some instances, the term **asphalt** is used instead of the term **bitumen** (Gruse and Stevens, 1960). In other instances, the term **asphalt** is used to indicate the natural or mechanical mixture of the organic binder having a considerable proportion of mineral matter present (Broome, 1973; Broome and Wadelin, 1973). In this respect, the term is applied to the organic/inorganic mixture that is used on roadways.

Using the conventions already applied (Chapter 1) will alleviate some of this confusion and it is this terminology that will be used here. Thus, for the purposes of this chapter, the term bitumen is used to designate the naturally occurring material,

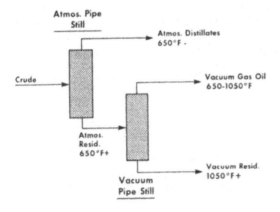

Figure 12.1: Crude oil distillation.

and the term **asphalt** is used to designate the petroleum product.

As noted (Chapter 1), and by way of a brief recap, a **residuum** (also shortened to "resid") is the residue obtained from petroleum after nondestructive distillation has removed all the volatile materials. The temperature of the distillation is usually maintained below 350°C (660°F), since the rate of thermal decomposition of petroleum constituents is substantial above 350°C (660°F).

Residua are black, viscous materials and are obtained by distillation of a crude oil under atmospheric pressure (atmospheric residuum) or under reduced pressure (vacuum residuum) (Figure 12.1). They may be liquid at room temperature (generally atmospheric residua) or almost solid (generally vacuum residua), depending upon the nature of the crude oil.

On the basis that petroleum is a continuum (Figure 12.2) (van Nes and van Westen, 1951; Gruse and Stevens, 1960; Speight, 1991), and no matter what the crude oil "type," the character of the residuum is determined by the character of the parent crude oil. In terms of crude oil "mapping" by fractionation studies (Speight, 1991), a residuum can also be further defined as a mixture of polar constituents (asphaltenes and resins), saturates, and aromatics (Figure 12.3) the relative amounts of the saturates depending upon the nature of the crude oil and the "cut point" of the residuum.

For example, using another form of crude oil "map" in which

Table 12.1: Different forms of asphalt

Natural Asphalt (Bitumen)	Manufactured Asphalt
Lake asphalt	Straight-run asphalt
Rock asphalt	Propane asphalt
Tar sand bitumen	Air-blown asphalt
	Thermal asphalt (pitch*, tar*)
	Blended asphalt
	Emulsion asphalt

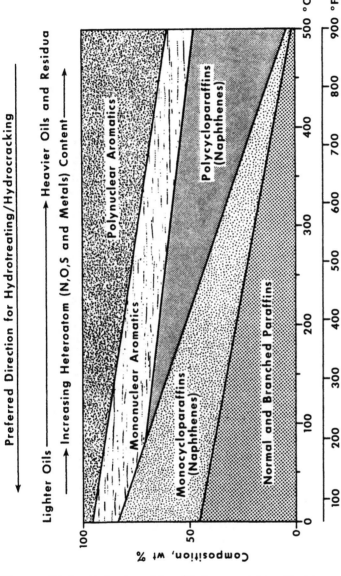

Figure 12.2: Hydrotreating/Hydrocracking Continuum for Petroleum.

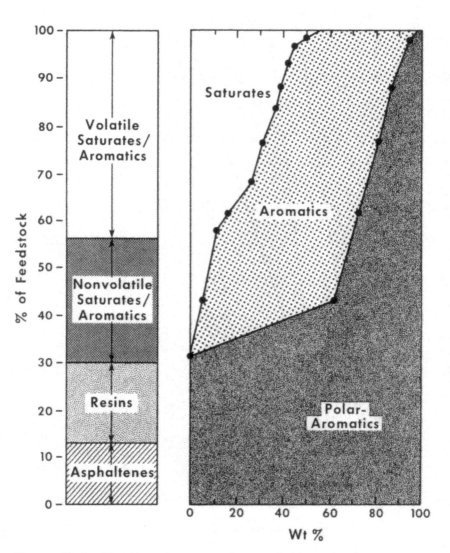

Figure 12.3: The Constituents of a Typical Residuum.

the boiling range and carbon number are employed, it is possible to define the boundaries for atmospheric residua and for vacuum residua (Figure 12.4) as well as for heavy oil (Figure 12.5).

To add further to a discussion of properties, one of the most important nonvolatile products is asphalt. Properties of asphalt are dictated by composition, which by extension and inference, determine the stability/compatibility of the materials with other asphalts.

A crude oil having a paraffinic base would yield a residuum of similar character, whilst a crude oil having a naphthenic/aromatic base would yield a residuum of a naphthenic character. The former (i.e., the paraffin base residuum) would yield an asphalt of inferior rheological, properties whereas a naphthenic/aromatic base would yield an asphalt having superior rheological properties.

Asphalt obtained by vacuum processes alone would exhibit lower penetration indices, i.e., a sol structure. Asphalt obtained by air-blowing vacuum residua (oxidation processes) would be characterized by higher penetration indices, indicating a more structured gel structure. Asphalt produced by propane deasphalting (Chapter 5) would have a very low penetration (hard asphalt) and when added to vacuum residua would impart satisfactory rheological properties (Gruse and Stevens, 1960).

When a residuum is obtained from a crude oil and thermal decomposition has commenced, it is more usual to refer to this product as **pitch**. The differences between a parent petroleum and the residua are due to the relative amounts of various constituents present (Figure 12.6) (Chapter 1), which are removed or remain by virtue of their relative volatility.

The chemical composition of a residuum from an asphaltic crude oil is complex (Figure 12.7). Physical methods of fractionation usually indicate high proportions of asphaltenes and resins in the residuum, even in amounts up to 50% (or higher). In addition, the presence of ash-forming metallic constituents, including such organometallic compounds as those of vanadium and nickel, is also a distinguishing feature of residua and the heavier oils. Furthermore, the deeper the "cut" into the petroleum, the greater is the concentration of sulfur and metals in the residuum, and the greater the deterioration in physical properties.

When the asphalt is produced simply by distillation of an

307

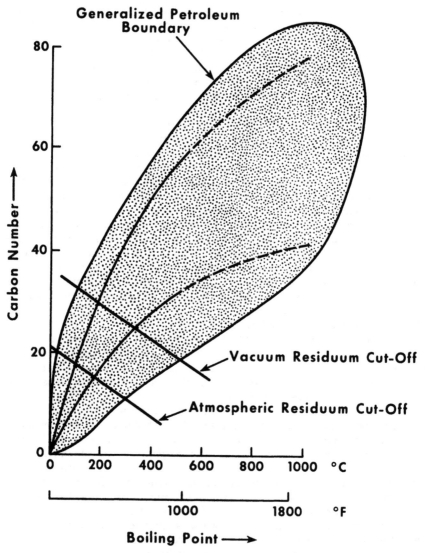

Figure 12.4: The Boiling Range and Carbon Number Boundaries for Atmospheric and Vacuum Residuum.

308

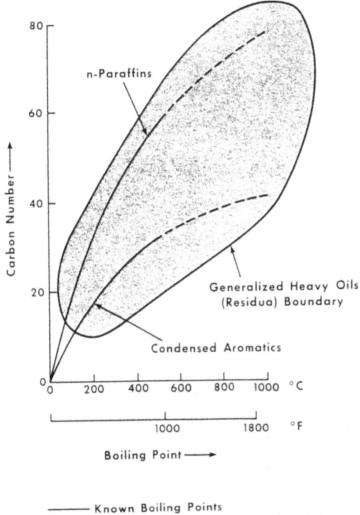

Figure 12.5: The Boiling Range and Carbon Number Boundaries for Heavy Oils and Residua.

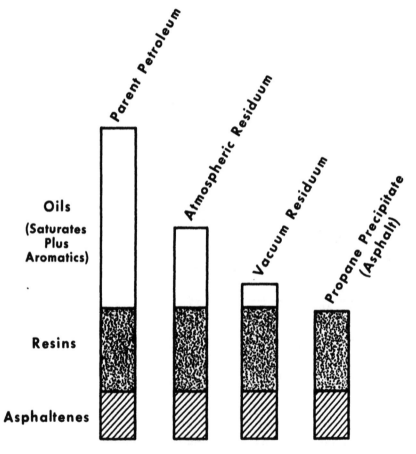

Figure 12.6: The Differences Between a Parent Petroleum and Residua.

asphaltic crude, the product can be referred to as residual, or straight-run, petroleum asphalt. If the asphalt is prepared by solvent extraction of residua or by light hydrocarbon (e.g., propane, butane) deasphalting, or if blown or otherwise treated, the term should be modified accordingly to qualify the product (e.g., propane asphalt).

By reference, bitumen is frequently found filling pores and crevices of sandstones, limestones, or argillaceous sediments, in which case the organic and associated mineral matrix is known as rock asphalt. The term bitumen includes a wide variety of reddish brown to black materials of semisolid, viscous to brittle character that can exist in nature with no mineral impurity or with mineral matter contents that exceed 50% w/w.

Tar is the result of the destructive distillation of many bituminous or other organic materials and are brown to black, oily, viscous liquids.

Tar is produced from coal and is generally understood to refer to the coal product, although it is advisable to specify coal tar if there is the possibility of ambiguity when tar is generated from other organic materials.

The most important factor in determining the yield and character of the coal tar is the carbonizing temperature. Three general temperature ranges are recognized, and the products have acquired the designations low-temperature tar (ca. 450-700°C; 840-1290°F); mid-temperature tar (ca. 700-900°C; 1290-1650°F); and high-temperature tar (ca. 900-1290°C; 1650-2190°F). Tar released during the early stages of the thermal decomposition of coal (or other organic material) is called primary tar, since it represents a product that has been recovered without the secondary alteration that results from prolonged residence of the vapor in the heated zone.

Treatment of the distillate (boiling up to 250°C, 480°F) from the tar with caustic soda causes separation of a fraction known as tar acids; acid treatment of the distillate produces a variety of organic nitrogen compounds known as tar bases. The residue left following removal of the heavy oil, or distillate, is pitch, a black, hard, and highly ductile material.

Although heavy oil can resemble **bitumen,** the definition of **heavy oil** is usually based on the API gravity or viscosity, and the definition is quite arbitrary, although there have been attempts to rationalize the definition based upon viscosity, API gravity, and density.

Thus, the generic term **heavy oil** is often applied to petroleum that has an API gravity of less than 20° and usually, but not always, a sulfur content higher than 2% w/w. Furthermore, in contrast to conventional crude oils, heavy oils are darker in color and may even be black.

The term **heavy oil** has also been arbitrarily used to describe both the heavy oils that require thermal stimulation of recovery from the reservoir and the bitumen in bituminous sand (tar sand) formations from which the heavy bituminous material is recovered by a mining operation.

2.0 Properties

Asphalt is a complex mixture of hydrocarbon and heteroatom constituents (Abraham, 1945; Gruse and Stevens, 1960; Barth, 1962; Hanson, 1964; Speight, 1991). The complex composition is reflected in the properties and behavior of asphalt (Speight, 1992).

The properties of any of the aforementioned materials/products can change during storage. However, it is more pertinent at this time, in keeping with the trend of the previous chapters, to consider the changes that occur in asphalt.

Asphalt can change during storage, resulting in degradation of the product. The degradation of asphalt can be estimated from changes in the penetration (ASTM D-5) or from changes in viscosity (ASTM D-2170). A penetration drop below the required limit or a viscosity rise above the specification limit may disqualify the asphalt for use.

In addition to these two properties, two additional rheological estimates are included in most of the asphalt specifications: the Ring and Ball softening point (ASTM D-36) and the ductility (ASTM D-113).

All four properties are basic rheological properties, determining the classification and the quality of asphalt. At the same time, all of them are subject to changes during storage and handling and are affected by environmental conditions: temperature, exposure to air (oxygen), and atmospheric radiation.

On the basis of this observation, a test is in current use indicating the rate of change of penetration, viscosity, and

ductility in a film of asphalt   heated in an oven for 5 hours at
163°C (325°F) on a rotating plate.

This test, the thin film oven test (TFOT) (ASTM D-1754),
establishes the effects of heat and air on the basis of changes
incurred in the above physical properties measured before and after
the oven test. The allowed rate of changes in the relevant asphalt
properties, after the exposure of the tested sample to the oven
test, are specified in the relevant specifications (ASTM D-3381).

Some of the other inspection data relating to asphalt quality
requirements are:

(1) the Cleveland open cup flash point (ASTM D-92), a safety
requirement;

(2) solubility in trichloroethylene (ASTM D-1981), an estimate
pertaining to the required absence in the asphalt of insoluble
impurities;

(3) the Fraas breaking point (IP 80) an estimate of
brittleness expressed as the minimal temperature at which a
thin layer of asphalt on a metal plate will not crack when it
is bent at specified conditions; and

(4) the asphaltene content (IP 143), an indication as to the
chemical nature of the asphalt.

3.0 Stability and Incompatibility

The stability of asphalt is associated with what is called
durability in asphalt paving technology. The former relates to
changes with time in the original properties of the asphalt in
general, while the latter refers rather to an age hardening
process, but both these phenomena are related to each other.

Asphalt produced by vacuum distillation tends to exhibit
instability, manifested by a drop in the penetration and ductility
and an increase in viscosity during storage and handling. The TFOT
(ASTM D-1754) is used as an estimate for this characteristic.

The term **long term stability** refers to the durability of
asphalt not only in its original form, but also in the form of
asphalt paving cements.

After application of the asphalt in road pavements, it is

exposed to extremes of environmental conditions: high temperatures (especially in warm regions), subzero temperatures in near-arctic regions, atmospheric radiation, and mechanical stress. Rheological as well as chemical considerations, have therefore to be taken into account.

Asphalt is characterized by an extremely complex chemical system of high-molecular-weight hydrocarbons, predominantly of a cyclic nature with paraffinic side chains and containing oxygen, sulfur and nitrogen components in their structure.

Asphalt is basically a colloidal solution in which the dispersed phase are the asphaltenes, or rather the asphaltene micelles, and the dispersing phase contains the maltenes.

The asphaltenes are defined as components insoluble in paraffinic solvents as n-pentane or n-heptane, but soluble in aromatic solvents; maltenes are defined as components soluble in paraffinic solvents (Figure 12.7). Carbenes are insoluble in aromatic solvents but are soluble in carbon tetrachloride or trichloroethylene. Carboids are insoluble in all of the solvents that dissolve asphaltenes and carbenes.

The asphaltene micelles are composed of an asphaltene core surrounded by resins and high molecular weight aromatic compounds, passing into a media of lower molecular weight aromatic compounds as an intermediate phase which is dispersed in the dispersing phase of the mineral oils (the so-called maltenes).

The chemical nature of asphalt, which is influenced by the production methods, affects its behavior in this respect. And many of the general properties of asphalt, which also serve as methods for routine inspection, are actually stability estimates.

Many asphalt properties change during storage and handling. Some asphalts exhibit a drop in penetration; for example, an asphalt might suffer a drop in penetration from 65 to 58 in a period of a few days and so turn from a specification product to a non-acceptable product.

Although not defined as a stability property of asphalt (since it measures decreases in penetration and ductility and increases in viscosity) a thin layer of which is exposed to heat and air. This exposure to heat and air of a thin film of an unstable asphalt would support formation of oxygen-containing polymerization products, which, in turn, would reduce the penetration of the

314

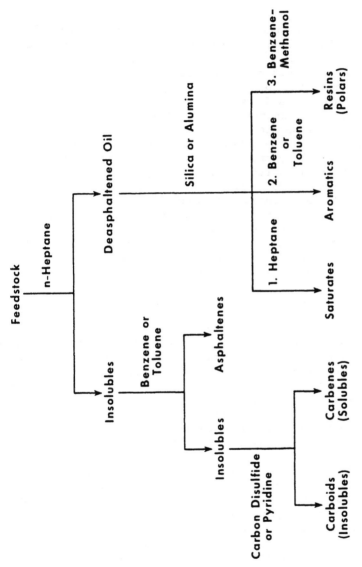

Figure 12.7: Chemical Fractionation and Composition.

asphalt (increase its hardness), reduce its ductility (make it more brittle or less elastic) and increase its viscosity. A stable asphalt would be affected much less by this exposure to heat and oxygen. There is a direct relation between the results of this test and the changes in the original properties of the asphalt during handling, storage and use.

Since asphalt is considered, with some justification, as a colloidal system, the nature of this system will determine the rheological properties of the asphalt, defined by its penetration, softening point, ductility and viscosity at given temperatures.

At this point it is worth considering again (see Chapter 11) the approach applied to the instability/incompatibility of heavy residual fuels and its potential application to asphalt.

This concept bases the instability/incompatibility on the chemical composition as well as on the internal colloidal structure (Por, 1992) by defining a colloidal instability index (CII). This is defined as the ratio of the sum of asphaltenes and saturated oils to the sum of the resins and aromatic solvents:

CII = (Asphaltenes + Saturates)/(Aromatics + Resins)

The equilibrium of a well peptized asphaltene system, such as asphalt, can be easily disturbed by (1) application of heat during service on days of extremely high temperatures and/or friction from vehicle tires; (2) oxidation due to constant exposure to air; (3) ultraviolet irradiation by prolonged exposure to sunlight; and (4) addition of a paraffinic diluent.

In each case, the chemical composition is altered and the aromaticity is affected, thereby causing a disruption of the equilibrium of the colloidal system (Moschopedis and Speight, 1977, 1978).

As a result, asphaltene particles are deprived of their enveloping layers, which previously merged continuously with the consecutive layers. The micelle system becomes noncontinuous, and the asphaltene cores are prone to agglomeration. Such a process leads to instability of the asphalt, perhaps resulting in phase separation of the asphaltenes from the asphalt thereby causing a loss of the asphalt-binder interactions. Pavement failure is the result.

Another of the estimates regarding the rheological properties is the **penetration** index.

Since the logarithm of the penetration is in a linear relation to the temperature;

$$\log pen = AT + K$$

where A is the slope of the temperature sensitivity as related to the logarithm of the penetration.

By extrapolation to the softening point temperature, a penetration of approximately 800 is obtained. The slope, A, can be obtained by measurement of the penetration at two different temperatures or by the relation of the penetration and the softening point temperature.

A penetration index (PI) can be obtained by:

$$d \log pen/dt = (20 - PI)/(10 + PI) \text{ x } 0.02$$

$$A = d \log pen/dt$$

$$PI = 10 \text{ x}(2 - 50A)/(1 + 50A)$$

A penetration index above +2 would indicate a gel structure of elastic properties and a thixotropic nature. A penetration index of below -2 would indicate a sol structure of Newtonian properties, whereas an asphalt exhibiting satisfactory rheological properties should have a penetration index between +1 and -1.

After an initial deformation, with a certain elasticity, Newtonian properties should prevail, with a proportionality between the deformation rate and the applied stress. Curves above these regions indicate gel-structured asphalts, and curves below these regions indicate sol-structured asphalts.

The presence of propane-precipitated asphalt in the asphalt mix improves the stability properties of such blends because of both the rheological and the chemical nature of the propane-precipitated asphalt. The improved stability properties of such blends can be seen from the decreasing differences in the viscosities, penetrations, and ductilities after exposure to elevated temperatures and oxygen (as, for example, in the TFOT).

The beneficial effect of the propane-precipitated asphalts is,

317

of course, subject to limitations in their proportions; up to 35% of the propane-precipitated asphalt in blends with vacuum residua would be the upper limit, much depending on the nature of the vacuum residues as well as on the nature and proportions of other components, as for example, lubricating oil extracts, which are sometimes used in such blends (Ishai et al., 1988).

The indications are that long-term asphalt stability is related to basic rheological and physicochemical characteristics of original as well as aged asphalt samples of varying compositions. It is also believed (Ishai et al., 1988) that the understanding of the relations of these characteristics on the properties of asphalt (paving asphalt cement) makes possible the prediction of asphalt durability performance in the field. The stability is indicated in this case by aging indices: viscosity and softening point ratios and retained penetration percentage before and after exposure to the TFOT (Por, 1992).

4.0 References

Abraham, H. 1945. Asphalts and Allied Substances. Volume 1. Van Nostrand, New York. Chapter II.

Barth, E.J. 1962. Asphalt: Science and Technology. Gordon and Breach, New York.

Broome, D.C. 1973. In Modern Petroleum Technology. G.D. Hobson and W. Pohl (editors). Applied Science Publishers, Barking, Essex, England. Chapter 23.

Broome, D.C., and Wadelin, F.A. 1973. In Criteria for Quality of Petroleum Products. J.P. Allinson (editor). Halsted Press, Toronto. Chapter 13.

Gruse, W.A., and Stevens, D.R. 1960. Chemical Technology of Petroleum. McGraw-Hill, New York.

Hanson, W.E. 1964. In Bituminous Materials: Asphalts, Tars, and Pitches. Volume I. A.J. Hoiberg (editor). Interscience Publishers, New York. Chapter 1.

Ishai, I., Brule, B., Vaniscote, J.C., and Raymond, G. 1988. Proceedings, Association of Asphalt Paving Technologists.

Moschopedis, S.E., and Speight, J.G. 1977. J. Materials Sci. 12: 990.

Moschopedis, S.E., and Speight, J.G. 1978. Fuel. 57: 235.

Por, N. 1992. Stability Properties of Petroleum Products. The Israel Institute of Petroleum and Energy, Israel.

Speight, J.G. 1991. The Chemistry and Technology of Petroleum. 2nd Edition. Marcel Dekker Inc., New York.

Speight, J.G. 1992. Asphalt. In Kirk-Othmer's Encyclopedia of Chemical Technology. 3: 689.

van Nes, K., and van Westen, H.A. 1951. Aspects of the Constitution of Mineral Oils. Elsevier, Amsterdam.

CHAPTER 13: ASPHALTENES AND INCOMPATIBILITY

1.0 Introduction

In a mixture as complex as a residuum or heavy oil, refining chemistry can only be generalized because of difficulties in analyzing not only the products but also the feedstock and the intricate and complex nature of the molecules that make up the feedstock. Moreover, the incompatibility of feedstocks and products is an ongoing issue during refining. The occurrence of sediment during thermal operations and from products during storage reduces the efficiency of a variety of processes.

The stability of fuels has been well researched over the past decades (Mushrush et al., 1989; Pedley et al., 1989; Hardy and Wechter, 1990; Hazlett et al., 1991), and there have been inroads made into the chemistry of fuel stability or incompatibility. Moreover, it is the general consensus that it is the polar, i.e. heteroatom, constituents of feedstocks that are responsible for the formation of such sediments.

However, an area that remains largely undefined, insofar as the chemistry and physics are still speculative, is the phenomenon of incompatibility as it occurs in refinery operations, especially when the heavier feedstocks, containing high proportions of asphaltenes, are employed.

At this point, it should be noted that it is not the intent to produce a similar text on the structural entities in coal (Gorbaty and Ouchi, 1981; Schobert et al., 1991; Speight, 1994) or for that matter, of the kerogen in oil shale (Scouten, 1990). The production of liquids from petroleum, coal, and oil shale is a complex operation, further complicated by the nature of the high-molecular-weight organic species.

The complex nature of these species can be taken several steps further when it is acknowledged that aphaltenes, coal, and kerogen are multidimensional in space (Gorlov and Golovin, 1993; Krichko et al., 1993). This is in contrast to the two-dimensional aspects often represented, for convenience, "on paper."

Reference has been made in earlier chapters to the issue of incompatibility when asphaltenes are present in the feedstock. In fact, asphaltenes appear to play a major role in interactions that lead to incompatibility in feedstocks, particularly during refining operations.

PETROLEUM PRODUCTS: INSTABILITY AND INCOMPATIBILITY

A key to decreasing incompatibility during refinery operations is to develop an understanding of the asphaltenes, which are the principal precursors to sludge and other sediments.

Indeed, the presence of asphaltenes in feedstocks is of considerable concern to those processes that center around the use of a catalyst, which due to incompatibility, which can lead to the deposition of metals and carbonaceous residues onto the catalyst (Speight, 1991).

This chapter summarizes the data relating to asphaltene characteristics which can be used to explain various aspects of asphaltene behavior, including sediment deposition in petroleum reservoirs, sludge formation, and refining/conversion to fuel oil, asphalt, and the like.

2.0 Separation

Asphaltenes are, by definition, a solubility class that is precipitated from petroleum, heavy oil, and bitumen by the addition of an excess of liquid hydrocarbon. However, the nomenclature of the nonvolatile constituents of petroleum (i.e., the asphaltenes, the resins, and to some extent, part of the oils fraction insofar as nonvolatile oils do occur in residua and other heavy feedstocks) is an operational aid and is not usually based on chemical or structural features.

3.0 Composition

The asphaltene fraction of petroleum is a solubility class and, as such, is composed of many different types of molecular species (Speight, 1991). Therefore, there is no one description that can be used to define an asphaltene.

Analytical data from various techniques and methods can only produce average values; such data do show that the asphaltene fraction contains the most aromatic and most polar species originally present in the crude oil (Figure 13.1). However, caution must be used in the application of average data to structural studies, where the formula for an average molecule might be derived. An **average molecule** that might be derived from the data may be very far from the truth and may only explain asphaltene behavior with difficulty and with much imagination!

3.1 Ultimate (Elemental) Composition

The elemental compositions of asphaltenes vary over only a narrow range, corresponding to H/C ratios of 1.15 ± 0.05%, although

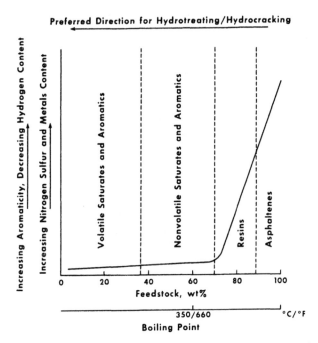

Figure 13.1: Illustration of heteroelement distribution and other properties of petroleum fractions.

values outside of this range are often found (Table 13.1) (Speight, 1991). Notable variations do occur in the proportions of the heteroelements, in particular, in the proportions of oxygen and sulfur. On the other hand, the nitrogen content of the asphaltenes has a somewhat lesser degree of variation. This is perhaps not surprising, since exposing asphaltenes to atmospheric oxygen can substantially alter the oxygen content, and exposing a crude oil to elemental sulfur, or even to sulfur-containing minerals, can result in sulfur uptake.

Table 13.1: General range of asphaltene composition.

GENERAL COMPOSITION

	%
Carbon	82.0 ± 3.0
Hydrogen	8.1 ± 0.7
Nitrogen	1.0 ± 0.5
Oxygen	1.0 ± 0.8
Sulfur	5.0 ± 3.0

ATOMIC RATIOS

H/C	1.15 ± 0.05
N/C	0.100 ± 0.003
O/C	0.020 ± 0.005
S/C	0.030 ± 0.005

3.2 Fractional Composition

Asphaltenes can be fractionated by the use of a variety of techniques (Bestougeff and Darmois, 1947, 1948; Altgelt, 1965; Bestougeff and Mouton, 1977; Francisco and Speight, 1984). Of specific interest is the observation that when asphaltenes are fractionated on the basis of aromaticity and polarity, it appears that the more aromatic species contain higher amounts of nitrogen (Figure 13.2) (Speight, 1984). This suggests that the nitrogen species are located predominantly in aromatic systems.

However, more important, the fractionation of asphaltenes into a variety of functional (and polar) types (Figure 13.3) (Francisco

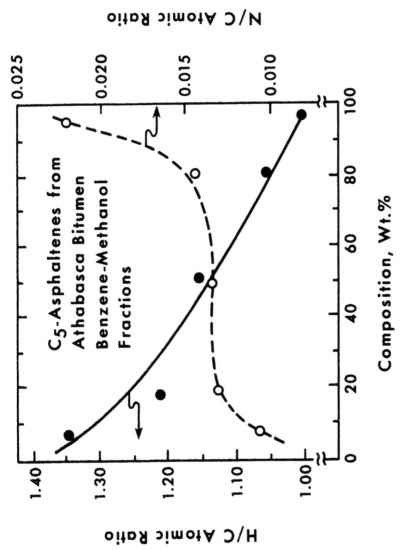

Figure 13.2: Fractionation of Asphaltenes and the Comparison of H/C to N/C Atomic Ratios.

and Speight, 1984) has confirmed the complexity of the asphaltene fraction. High performance liquid chromatography (HPLC) has also confirmed the diversity of the structural and functional types in asphaltenes to the extent that different HPLC profiles can be expected for different asphaltenes (Figure 13.4).

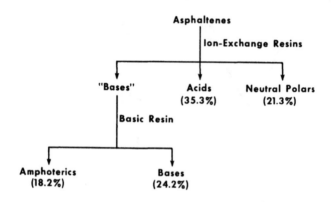

Figure 13.3: Asphaltenes subfractionated on the basis of functionality.

Coal asphaltenes (and, by inference, shale oil asphaltenes which are also a thermal degradation product) can also be placed on this scale on the basis that they are of lower molecular weight and more polar than petroleum asphaltenes. In addition, the molecular weight/polarity concept can also be used to indicate the relative nature of carbenes and carboids (thermal decomposition products of asphaltenes), which are of lower molecular weight and more polar than their precursors (Figure 13.5).

However, it should not be construed that this two-dimensional diagram is truly representative of the natures of asphaltenes (petroleum and coal) and carbenes and carboids. If it was, it might be concluded that coal asphaltenes and petroleum carbenes/carboids are similar molecular species. This is not believed to be the case, and a three-dimensional diagram would be more appropriate to show the chemical and physical differences between the various asphaltenes/carbenes/carboids.

In summary, fractionation data lend support to, and reinforce, the concept that asphaltenes are complex mixtures of molecular sizes and various functional types (Figure 13.6) (Long, 1979, 1981) and that carbenes and carboids can also fit into this concept (Figure 13.7).

Although coal asphaltenes have also been included on the molecular weight/polarity scale (Figure 13.8), it must not be presumed that they are the same as carbenes/carboids formed in thermal reactions from petroleum feedstocks. A three-dimensional plot involving functionality would be more appropriate. However, the concept does emphasize the lower molecular weight and increased polarity (relative to petroleum asphaltenes) of coal asphaltenes and carbenes/carboids.

4.0 Structural Studies

The molecular nature of the nonvolatile fractions of petroleum has been the subject of numerous investigations (Speight, 1972, 1991; Yen, 1972, 1974). The data from these studies intimated that asphaltenes, viewed structurally, contain condensed polynuclear aromatic ring systems bearing alkyl side chains.

It was generally believed that as the boiling point of the crude oil fraction increased, the size of the aromatic polynuclear systems increased, although (on the basis of fractionation studies) it would be more appropriate to place the asphaltenes in the less well defined region marked **polar aromatics** (Figure 13.9). This is generally true to a point but it is more than likely that the size of the condensed ring system is more in keeping with the skeletal structures of natural product ring systems, where the number of condensed rings very rarely exceed six. In fact, it is more likely that the resin and asphaltenes fraction contribution to the higher boiling point involves the inclusion of heteroatom systems, longer alkyl side chains, and naphthenic systems. This would be more consistent with the concept that petroleum is a continuum of chemical types

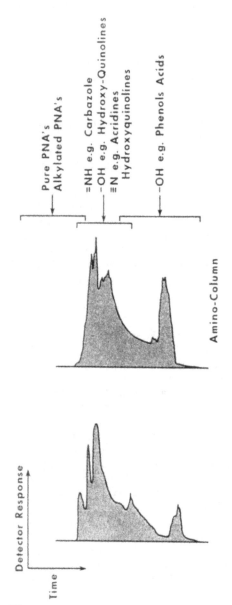

Figure 13.4: HPLC Profile of Asphaltenes.

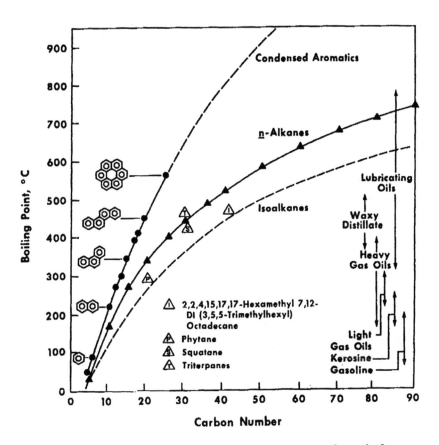

Figure 13.5: Thermal Decomposition Products of Asphaltenes.

Thus, with the conception of larger polynuclear aromatic systems in mind, it is not surprising that the formulae derived from the studies invoked the concept of large polynuclear aromatic systems and efforts were made to describe the total structures of asphaltenes (Figure 13.10) in accordance with magnetic resonance data and results of other spectroscopic and analytical techniques. The macrostructure of the asphaltenes was believed to involve graphite-like stacks (Figure 13.11).

Nitrogen occurs in asphaltenes as various heterocyclic types (Speight, 1991 and references cited therein). The more conventional types of nitrogen, i.e., primary, secondary, and tertiary amines, have not been established as being present in petroleum asphaltenes. The nitrogen has also been defined in terms of basic and nonbasic types (Nicksic and Jeffries-Harris, 1968).

Oxygen has been identified in carboxylic, phenolic, and ketonic locations (Nicksic and Jeffries-Harris, 1968; Petersen et al., 1974; Speight and Moschopedis, 1981).

Sulfur occurs in thiane, thiolane, thiophene, benzothiophene,

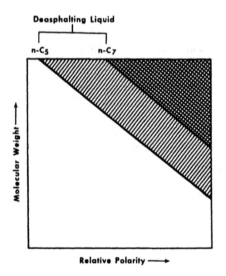

Figure 13.6: Asphaltenes are a combination of various types

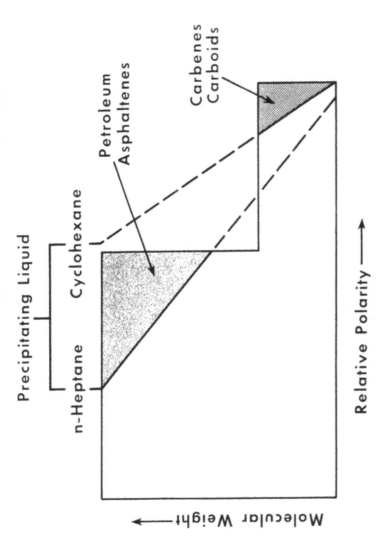

Figure 13.7: Carbenes and Carboids in Asphaltene Fractionation.

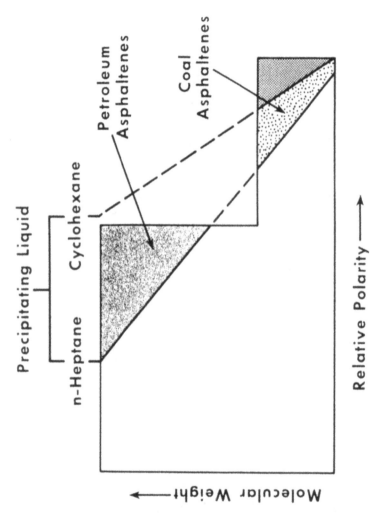

Figure 13.8: Petroleum and Coal Asphaltenes Compared on a Molecular Weight/Polarity Scale.

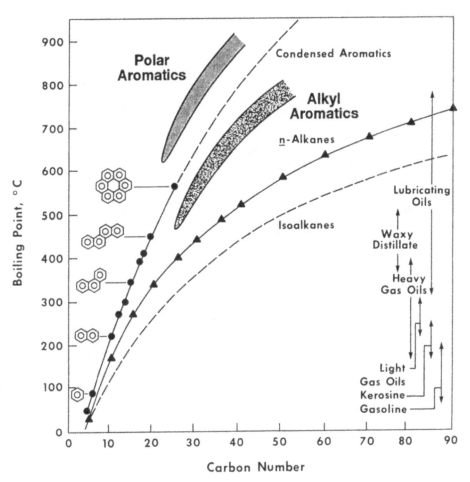

Figure 13.9: Polyaromatic Systems and Boiling Point.

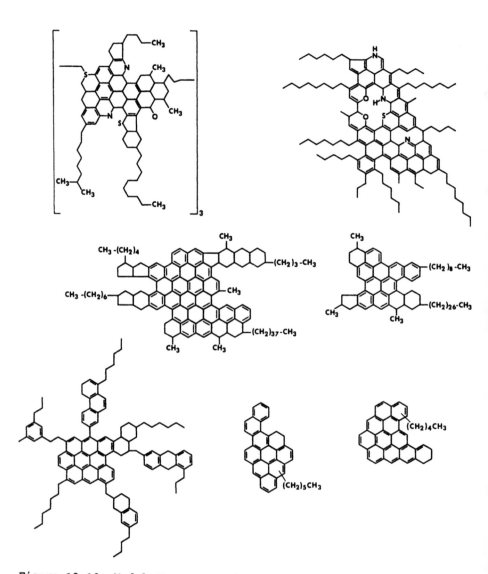

Figure 13.10: Model Structure of Asphaltenes.

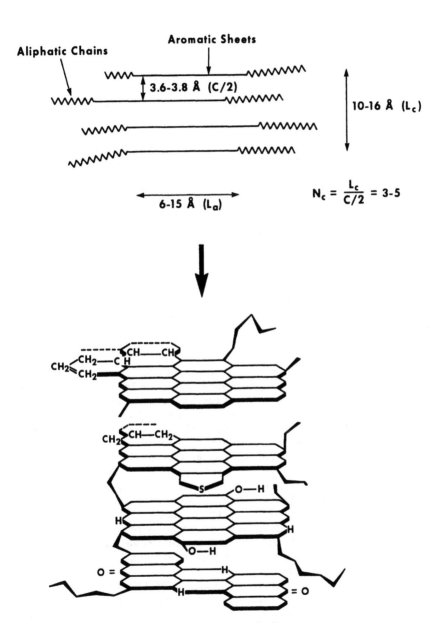

Figure 13.11: Macrostructure of Asphaltenes.

dibenzothiophene, and naphthenobenzothiophene as well as in systems of alkyl-alkyl sulfides, alkyl-aryl sulfides, and aryl-aryl sulfides (Yen, 1974; Speight and Pancirov, 1984). More highly condensed thiophene-types may also exist but are precluded from identification by low volatility. Investigations of the size distribution of the sulfur species indicate an increase in the size of the sulfur species from aromatics to resins to asphaltenes.

Metals (i.e., nickel and vanadium) are much more difficult to integrate into the asphaltene system. The nickel and vanadium occur as porphyrin systems (Baker, 1969; Yen, 1975), but whether or not these are an integral part of the asphaltene structure is not known. Some of the porphyrin systems can be isolated as a separate stream from petroleum (Branthaver, 1990).

5.0 Molecular Weight
Asphaltene molecular weights are variable (Yen, 1974; Speight et al., 1985), having the tendency to associate even in dilute solution in nonpolar solvents. However, data produced using highly polar solvents indicate that the molecular weights, in solvents that prevent association, usually fall into the range 2000 ± 500.

6.0 Molecular Models
The overall goal of structural studies should not be merely derivation of a model for asphaltenes but should be use of the model to predict the behavioral characteristics during processing. Predictability, in addition to rationalization of process chemistry, then become useful functions of the model and are reasons for the existence of the model.

In terms of the derivation of the model, it is necessary that models derived for the higher-molecular-weight constituents be consistent with the molecular species that are present in the lower-boiling fractions of petroleum. Inconsistency here would be a contradiction of the natural product origins of petroleum.

Thus, every attempt should also be made to be consistent with the chemical and physical effects that are prevalent in nature. The molecular types should be consistent with the skeletal structures of natural products.

The potential for skeletal alteration of the natural product systems, which originally occurred in the source material during geologic time and under geologic conditions, is more than likely, and pathways have been suggested (Tissot and Welte, 1978). In keeping with the natural product origins of petroleum, it is also assumed that the polynuclear aromatic systems in asphaltenes are based on the angular, phenanthrene, system in preference to the linear, anthracene, system (Fieser and Fieser, 1949).

Indeed, the concept of an asphaltene model that uses smaller polynuclear aromatic systems would be more in keeping with the types of polynuclear aromatic systems that occur in nature and the concept that petroleum is a continuum (Speight, 1991), thereby allowing the occurrence of smaller polynuclear aromatic systems throughout the various fractions.

One area where the large polynuclear aromatic systems have been assumed to be essential in the models is in the formation of high yields of coke from the asphaltenes. However, in support of the formation of a nonvolatile residue (coke) being formed from small condensed aromatic ring systems, it is worthy of note that heat-resistant polymers containing smaller aromatic and heterocyclic units such as polyquinolines and polyquinoxalines have a strong tendency to form large condensed systems during pyrolysis, which then carbonize (Dussel et al., 1982; NASA, 1994).

Thus, it may not be the mean size of the molecular entities that is responsible for the formation of coke, but there may also be structural or compositional features that contribute to coke formation.

Asphaltene models that might be more in keeping with behavioral characteristics might well be selected from the two extremes of the molecular weight/polarity diagram (Long, 1979, 1981) and would have smaller polynuclear aromatic systems and also span the range of functional types as well as molecular sizes. It must also be recognized that such models can have the large size dimensions that have been proposed for asphaltenes but will also be "molecular chameleons" insofar as they can vary in dimensions, depending upon the angle of rotation about an axis and/or the freedom of rotation about one, or more, of the bonds.

However, in the interests of caution, as advocated earlier in this chapter, it is necessary to reemphasize that, for the present purposes, these models are mere representations of molecular configurations and are used here for discussion only.

7.0 Asphaltenes in Petroleum

An early hypothesis of the physical structure of petroleum (Pfeiffer and Saal, 1940) indicated that asphaltenes are the centers of micelles formed by adsorption, or even by absorption of part of the maltenes, that is, resin material, onto the surfaces or into the interiors of the asphaltene particles.

Thus, most of those substances with greater molecular weight and with the most pronounced aromatic nature are situated closest to the nucleus and are surrounded by lighter constituents of a less aromatic nature. The transition of the intermicellar (dispersed or oil) phase is gradual and almost continuous (Figure 13.12). Since asphaltenes are incompatible with the oil fraction (Swanson, 1942; Koots and Speight, 1975), asphaltene dispersion is mainly attributable to the resins (polar aromatics).

On this basis, the stability of petroleum can be represented by a three-phase system in which the asphaltenes, the aromatics (including the resins), and the saturates are delicately balanced (Figure 13.13). Various factors, such as oxidation, can have an adverse effect on the system, leading to instability/incompatibility as a result of changing the polarity and bonding arrangements of the species in crude oil.

The evolution of this concept is perhaps the single most responsible agent for the perception that asphaltenes in petroleum exist in the manner of a colloidal system. Indices have been derived to describe this behavior (Chapter 12). One strong factor in favor of such a system is the inability of most crude oils to disperse the asphaltene fraction in the absence of the resin fraction (Koots and Speight, 1975).

It is some of these factors that are the focus of refining operations (Chapter 14) and that can lead to excessive coke formation during thermal processes.

Thus, it is possible to represent asphaltenes as smaller-ring (rather than graphitic-type) systems, where the skeletal structure is based on the carbon skeletons of various natural product systems (Figures 13.14 and 13.15). Furthermore, it is also possible to use the polarity/molecular weight concept to maximum advantage in such systems. Other molecular permutations are possible and it is not the intent here to claim that the structures in Figures 13.14 and 13.15 are the true structures of asphaltenes. It will, however, be shown that such structures can be employed to model the thermal

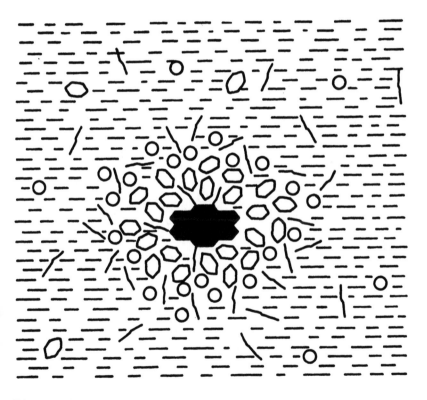

Figure 13.12: The Gradual Transition of the Intermicellar Phase.

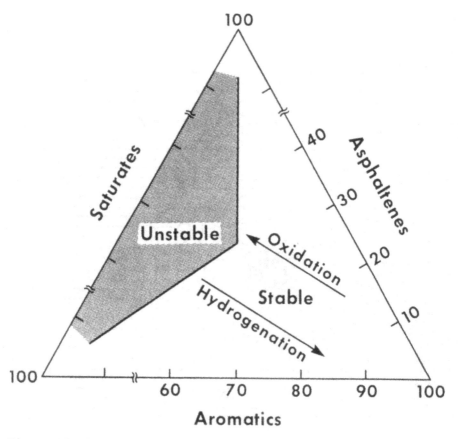

Figure 13.13: A Three Phase System Representing the Stability of Petroleum.

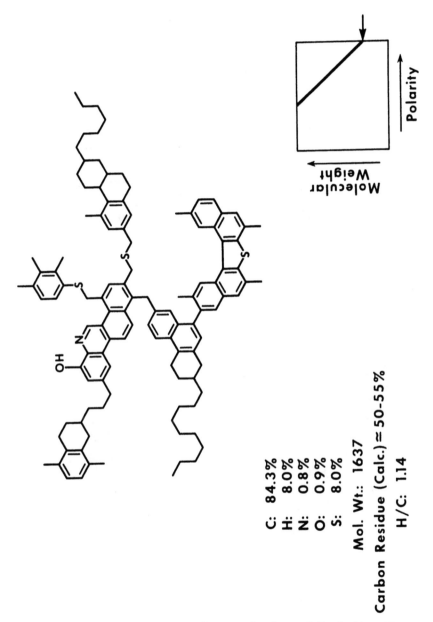

C: 84.3%
H: 8.0%
N: 0.8%
O: 0.9%
S: 8.0%

Mol. Wt.: 1637

Carbon Residue (Calc.) ≈ 50-55%

H/C: 1.14

Figure 13.14: The Smaller-Ring Depiction of Asphaltenes.

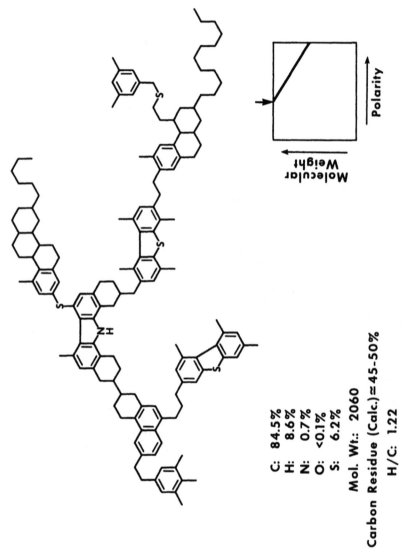

C: 84.5%
H: 8.6%
N: 0.7%
O: <0.1%
S: 6.2%

Mol. Wt.: 2060

Carbon Residue (Calc.) ≈ 45-50%

H/C: 1.22

Figure 13.15: Asphaltene Depicted as a Collection of the Skeletal Structures of Natural Products.

decomposition of asphaltenes and illustrate the phenomena of instability and incompatibility (Chapter 14).

8.0 Asphaltenes and Thermal Processes

The formation of solid sediments, or coke, during thermal processes is a major limitation to the process. Furthermore, the presence of these different types of solids shows that solubility controls solids formation. The tendency toward solids formation changes in response to the relative amounts of the light ends, middle distillates, and residues, and also in response to their changing chemical composition as the process proceeds (Gray, 1994).

The prime mover in the formation of incompatible products during the processing of feedstocks containing asphaltenes is the nature of the primary thermal decomposition products (Speight, 1987, 1992; Wiehe, 1992, 1993; Calemma et al., 1994).

There have been several attempts to focus attention on the asphaltenes during hydroprocessing studies. The focus has been on the macromolecular changes that occur by investigation of the changes to the generic fractions, i.e., the asphaltenes, the resins, and the other fractions that make up such a feedstock (Drushel, 1983).

This option suggests that the overall pathway by which hydrotreating and hydrocracking of heavy oils and residua occurs involves a stepwise mechanism:

Asphaltenes -------> polar aromatics (resin-type components)

Polar aromatics -------> aromatics

Aromatics -------> saturates

A direct step from either the asphaltenes or the resin-type components to the saturates is not considered a predominant pathway for hydroprocessing.

The means by which asphaltenes are desulfurized, as one step of a hydrocracking operation, are also suggested as part of this process. This concept can then be taken one step further to show the dealkylation of the aromatic systems as a definitive step in the hydrocracking process (Speight, 1987) (see also Chapter 14).

It is also likely that molecular species within the asphaltene

343

fraction that contain nitrogen and other heteroatoms, and have lower volatility than their hydrocarbon analogs, are the prime movers in the production of coke (Speight, 1987) (Chapter 14).

When catalytic processes are employed, complex molecules (such as those that may be found in the original asphaltene fraction or those formed during the process) are not sufficiently mobile (or are too strongly adsorbed by the catalyst) to be saturated by the hydrogenation components and, hence, continue to condense and eventually degrade to coke. These deposits deactivate the catalyst sites and eventually interfere with the hydroprocess.

A very convenient means of understanding the influence of feedstock on the hydrocracking process is through a study of the hydrogen content (H/C atomic ratio) and molecular weight (carbon number) of the feedstocks and products (O'Rear et al., 1982). Such data show the extent to which the carbon number must be reduced and/or the relative amount of hydrogen that must be added to generate the desired lower-molecular-weight, hydrogenated products.

In addition, it is also possible to use data for hydrogen in residuum processing, where the relative amount of hydrogen consumed in the process can be shown to be dependent upon the sulfur content of the feedstock (Bridge et al., 1975).

In summary, and as these concepts show, characterization data can be used as an integral part of refinery operations. There is the need to understand the nature of the asphaltenes and polar constituents of feedstocks in more detail in order to be able to predict product yield, product distribution, and product quality. It is this latter item that will allow some measure of predictability in terms of the compatibility/incompatibility of the the product mix. The characterization techniques currently at hand are an aid to accomplishing this goal.

9.0 References

Altgelt, K.H. 1965. J. Appl. Polymer Sci. 9: 3389.
Andersen, S.I., and Birdi, K.S. 1990. Fuel Sci. Technol. Int. 8: 593.
Baker, E.W. 1969. In Organic Geochemistry. G. Eglinton and M.T.J. Murphy (editors). Springer-Verlag, New York.
Bestougeff, M.A., and Darmois, R. 1947. Comptes Rend. 224: 1365.
Bestougeff, M.A., and Darmois, R. 1948. Comptes Rend. 227: 129.
Bestougeff, M. A., and Mouton, Y. 1977. Bull. Liason Lab. Pont. Chaus. Spec. Vol.: 79.
Branthaver, J.F. 1990. In Fuel Science and Technology Handbook. J.G. Speight (editor). Marcel Dekker Inc., New York.
Bridge, A.G, Scott, J.W., and Reed, E.M. 1975. Oil Gas J. 73(20): 94.
Calemma, V., Montanari, L., Nali, M., and Anelli, M. 1994 Preprints Div. Pet. Chem. Am. Chem. Soc. 39(3): 452.
Drushel, H.V. 1983. J. Chromatogr. Sci. 21: 375.
Dussel, H.J., Recca, A., Kolb, J., Hummel, D.O., and Stille, J.K. 1982. J. Anal. Appl. Pyrolysis. 3: 307.
Fieser, L.F., and Fieser, M. 1949. Natural Products Related to Phenanthrene. Reinhold Publishing Corp., New York.
Francisco, M. A., and Speight, J.G. 1984. Preprints Div. Fuel. Chem. Am. Chem. Soc. 29(1): 36.
Girdler, R.B. 1965. Proc. Assoc. Asphalt Paving Technol. 34: 45.
Gorbaty, M.L., and Ouchi, K. (Editors) 1981. Coal Structure. Advances in Chemistry Series No. 192. American Chemical Society, Washington, DC.
Gorlov, E.G., and Golovin, G.S. 1993. Khim. Tverd. Topl. No. 6: 81.
Gray, M.R. 1994. Upgrading Petroleum Residues and Heavy Oils. Marcel Dekker Inc., New York.
Hardy, D.R., and Wechter, M.A. 1990. Fuel. 69: 720.
Hazlett, R.N., Schreifels, J.A., Stalick, W.M., Morris, R.E., and Mushrush, G.W. 1991. Energy Fuels. 5: 269.
Koots, J.A., and Speight, J.G. 1975. Fuel. 54: 179.
Krichko, A.A., Gagarir, S.G., and Maker'ev, S.S. 1993. Khim. Tverd. Topl. No. 6: 21.
Long, R.B. 1979. Preprints Div. Pet. Chem. Am. Chem. Soc. 24(4): 891.

Long, R.B. 1981. In The Chemistry of Asphaltenes. J.W. Bunger and N. Li (editors). Advances in Chemistry Series No. 195. American Chemical Society, Washington, DC.

Mitchell, D.L., and Speight, J.G. 1973. Fuel. 52: 149.

Mushrush, G.W., Watkins, J.M., Beal, E.J., Morris, R.E., Cooney, J.V., and Hazlett, R.N. 1989. Fuel Sci. Technol. Int. 7: 931.

NASA. 1994. Tech. Briefs. 18(2): 56.

Nicksic, S.W., and Jeffries-Harris, M.J. 1968. J. Inst. Pet. London. 54: 107.

O'Rear, D.J., Frumkin, H.A., and Sullivan, R.F. 1982. Proceedings, 47th Meeting, American Petroleum Institute, New York, May.

Pedley, J.F., Hiley, R.W., and Hancock, R.A. 1989. Fuel. 68: 27.

Petersen, J.C., Barbour, F.A., and Dorrence, S.M. 1974. Proc. Assoc. Asphalt Paving Technol. 43: 162.

Pfeiffer, J.P., and Saal, R.N. 1940. Phys. Chem. 44: 139.

Schobert, H.H., Bartle, K.D., and Lynch, L.J. (editors). 1991. Coal Science II. Symposium Series No. 461. American Chemical Society, Washington, DC.

Scouten, C. 1990. In Fuel Science and Technology Handbook. J.G. Speight (editor). Marcel Dekker Inc., New York.

Speight, J.G. 1972. Appl. Spectrosc. Rev. 5: 211.

Speight, J.G. 1984. In Characterization of Heavy Crude Oils and Petroleum Residues. B. Tissot (editor). Editions Technip, Paris.

Speight, J.G. 1987. Preprints Div. Pet. Chem. Am. Chem. Soc. Div. Petrol. Chem. 32(2): 413.

Speight, J.G. 1991. The Chemistry and Technology of Petroleum. 2nd Edition. Marcel Dekker Inc., New York.

Speight, J.G. 1992. Proceedings, 4th International Conference on the Stability and Handling of Liquid Fuels. Report DOE/CONF-911102. U.S. Department of Energy, Washington, DC. p. 169.

Speight, J.G. 1994. The Chemistry and Technology of Coal. 2nd Edition. Marcel Dekker Inc., New York.

Speight, J.G., and Moschopedis, S. E. 1981. Preprints. Div. Petrol. Chem. Am. Chem. Soc. 26(4): 907.

Speight, J.G., and Pancirov, R.J. 1984. Liquid Fuels Technol. 2: 287.

Speight, J.G., Long, R.B., and Trowbridge, T.D. 1984. Preprints Div. Fuel Chem. Am. Chem. Soc. 27(3/4): 268.

Speight, J.G., Wernick, D.L., Gould, K.A., Overfield, R.E., Rao, B.M.L., and Savage, D.W. 1985. Rev. Inst. Fr. Petrole. 40: 51.

Swanson, J.M. 1942. J. Phys. Chem. 46: 141.

Tissot, B.P., and Welte, D.H. 1978. Petroleum Formation and Occurrence. Springer-Verlag, New York.

Wiehe, I.A. 1992. Ind. Eng. Chem. Res. 31: 530.

Wiehe, I.A. 1993. Ind. Eng. Chem. Res. 32: 2447.

Yen, T.F. 1972. Preprints Div. Pet. Chem. Am. Chem. Soc. 17(4): F102.

Yen, T.F. 1974. Energy Sources. 1: 447.

Yen, T.F. 1975. The Role of Trace Metals in Petroleum. Ann Arbor Science Publishers, Ann Arbor, Michigan.

CHAPTER 14: INCOMPATIBILITY IN REFINING OPERATIONS

1.0 Introduction

Crude oil is rarely used in its raw form, but instead must be processed into its various products (Figure 14.1). Aside from contaminant minerals such as sulfur and small amounts of trace metals, which are removed during refining, petroleum is composed of hydrocarbons, essentially varying combinations of carbon and hydrogen atoms; any hydrocarbon can be converted into any other, given the appropriate application of energy, chemistry, and technology. The smaller the molecule and the lower the ratio of carbon to hydrogen, the lighter the hydrocarbon, the lower the evaporation temperature, and usually, the more valuable the product.

A refinery is a complex network of vessels, equipment, and pipes. The total scheme can be divided into a number of unit processes. Refined products establish the order in which each refining unit will be introduced. Only one or two key product specifications are used to explain the purpose of each unit. Nevertheless, the choices among several types of units and the sizes of these units are complicated economic decisions. The trade-off among product types, quantity, and quality will be mentioned to the extent that they influence the choice of one kind of process unit over another.

Each refinery has its own range of preferred crude oil feedstock for which a desired distribution of products is obtained. The crude oil usually is identified by its source country, underground reservoir, or some distinguishing physical or chemical property. The three most frequently specified properties are density, chemical characterization, and sulfur content.

Every crude oil contains a mix of different hydrocarbons, and the two tasks of a refinery are to separate them out into usable products and to convert the less desirable hydrocarbons into more valuable ones. The tall metal towers that characterize petroleum refineries are distillation, or fractionating, towers.

Distillation is the primary method used to refine petroleum. When the heated crude oil is fed into the lower part of a tower, the lighter oil portions, or fractions, vaporize.

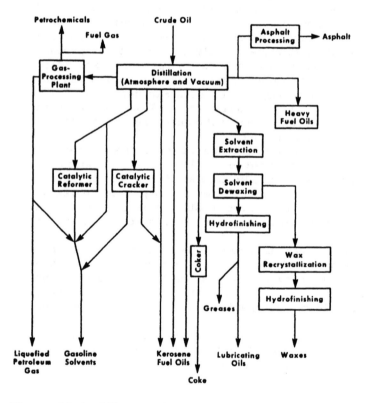

Figure 14.1: Illustration of a petroleum refinery.

Losing temperature as they rise, they condense into liquids, which flow downward into the higher temperatures and are revaporized. This process continues until the various fractions have achieved the appropriate degrees of purity.

The lighter fractions like butane, gasoline, and kerosene are tapped off from the top of the tower; heavier fractions like fuel and diesel oils are taken from the lower half. At more complex refineries the less valuable products of distillation are refined once again through various conversion processes, broadly referred to as "cracking."

Through the application of vacuums, heat, and catalysts, larger, heavier molecules are broken down into lighter ones. Thermal cracking, for instance, uses heat and pressure, while catalytic cracking employs a finely powdered catalyst, and hydrocracking involves the addition of hydrogen to produce compounds with lower carbon to hydrogen ratios, such as gasoline. Other processes produce high-octane products for blending with fuels, remove undesirable constituents, or make special petroleum compounds, including lubricants.

API gravity is a contrived measure of density that is related to the specific gravity:

$$API = (141.5/d) - 131.5$$

where d is the specific gravity, or the ratio of the weight of a given volume of oil to the weight of the same volume of water at a standard temperature, usually 15.6°C (60°F).

An oil with the same density as water, or with a specific gravity of 1.0, would then be 10° API oil. Oils with higher than 10° API gravity are lighter than water. Since lighter crude oil fractions are usually more valuable, a crude oil with a higher API gravity will bring a premium price in the market place.

Heavier crude oils are getting renewed attention as supplies of lighter crude oil dwindle. For convenience, heavy crude oil is defined as an oil with an API gravity of less than 20°.

A characterization factor was introduced by Watson and Nelson to use as an index of the chemical character of a crude oil or its fractions. The Watson characterization factor

PETROLEUM PRODUCTS: INSTABILITY AND INCOMPATIBILITY

(Chapter 7) is defined as follows:

$$\text{Watson } K = (T_B)^{1/3}/d$$

where T_B is the absolute boiling point in degrees Rankine and d is specific gravity at 15.6°C (60°F).

For a material with a wide boiling range like crude oil, the boiling point is taken as an average of the five temperatures at which 10%, 30%, 50%, 70%, and 90% is vaporized.

A highly paraffinic crude oil might have a characterization factor as high as 13, while a highly naphthenic crude oil could be as low as about 10.5. Highly paraffinic crude oils can also contain heavy waxes that make it difficult for the oil to flow. Thus, another test for paraffin content is to measure how cold a crude oil can be before it fails to flow under specific test conditions. The higher the pour point temperature, the greater the paraffin content for a given boiling range.

Once a refinery feedstock has been defined sufficiently well to indicate the primary methods of refining, it is then the nature of the processing operations that dictate the product slate. However, it is the nature of the feedstock and the products that dictate the potential for instability and/or incompatibility.

The potential for instability/incompatibility during refining operations is real and must be addressed during processing.

The purpose of this chapter is to indicate when and where the phenomena of instability/incompatibility can occur during refining operations as well as to present indications of mitigating these effects.

At this point it is worthy of note that there also exists the potential for synthetic fuels, because of the differences in composition and the more polar species produced with the hydrocarbons, to exhibit incompatibility when mixed with petroleum products and that some form of upgrading would be necessary (Figures 14.2 and 14.3).

Finally, the separation of solids during refining can occur during a variety of processes, either by intent (such as in the deasphalting process) or inadvertently, when the separation is detrimental to the process. Thus, separation of solids occurs

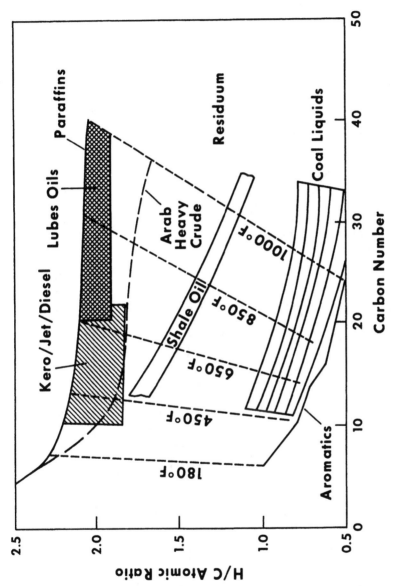

Figure 14.2: A Comparison of Carbon Number at H/C Ratio of Petroleum and Synthetic Fuels.

353

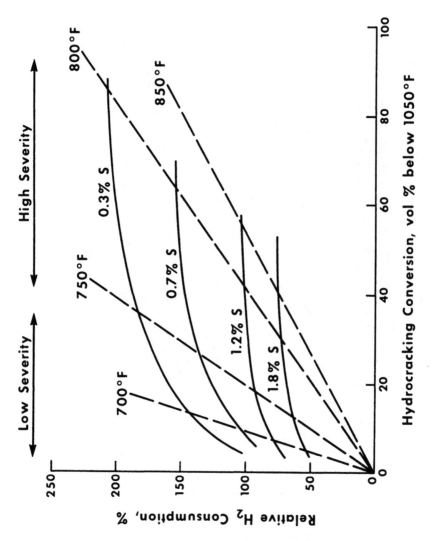

Figure 14.3: Hydrogen Consumption and Sulfur Content in Hydrocracking Conversion.

whenever the solvent characteristics of the liquid phase are no longer adequate to maintain polar and/or high molecular weight material in solution.

Examples of such occurrences are (1) asphaltene separation that occurs when the paraffinicity of the liquid medium increases; (2) wax separation that occurs when there is a drop in temperature or the aromaticity of the liquid medium increases; (3) sludge/sediment formation in a reactor that occurs when the solvent characteristics of the liquid medium change so that asphaltic or wax materials separate; (4) coke formation that occurs at high temperatures and commences when the solvent power of the liquid phase is not sufficient to maintain the coke precursors in solution; (5) sludge/sediment formation in fuel products that occurs because of the interplay of several chemical and physical factors (Chapter 6). This latter phenomenon will not be dealt with here, since the focus of this chapter is on the separation of insoluble phases during refining operations and not in the products.

2.0 Processes

When petroleum fractions are heated to temperatures over 350°C (660°F), the thermal or free radical reactions start to give way to cracking of the mixture at significant rates. Thermal conversion does not require the addition of a catalyst; therefore, this approach is the oldest technology available for residue conversion. The severity of thermal processing determines the conversion and the product characteristics.

The thermal treatment of residua ranges from mild, for reduction of viscosity, to ultrapyrolysis, for complete conversion to olefins and light ends (Speight, 1991; Gray, 1994). The higher the temperature, the shorter the time to achieve a given conversion. The severity of the process conditions is defined as the combination of reaction time and temperature to achieve a given conversion.

If no side reactions occur, then very long times at low temperature should be equivalent to very short times at high temperature. Thermal reactions, however, can give rise to a variety of different reactions, so that selectivity for a given product changes with temperature and pressure. Mild- and high-severity processes are frequently used for processing of residue fractions. More extreme conditions are only used commercially for cracking ethane, propane, butane, and light distillate feeds to produce ethylene and higher olefins.

The main limitation on straight thermal processing is that the products can be unstable. Thermal cracking at low pressure produces olefins, particularly in the naphtha fraction. These olefins give a very unstable product, which tends to undergo polymerization reactions to form tars and gums. The heavy fraction can form solids or sediments, which also feed composition, which also determines the maximum conversion allowable, ranging from 12% for South American crude through to 30% for North Sea atmospheric residue (Allan et al., 1983). However, before progressing to the thermal processes, there is one process in which incompatibility is used as a beneficial effect in product separation, and that is the deasphalting process.

2.1 Deasphalting

There are several processing operations worthy of note that are dependent upon the compatibility of asphaltenes and their products.

The chemistry of refining processes is often represented by simple chemical equations but is, in reality, a complex sequence of chemical reactions (Speight, 1991; Gray, 1994). The complexity of the feedstock influences the complexity of the products. Thus, there are many unknowns in the chemistry and physics of incompatibility during refining operations.

First, it must be noted that there are two general routes to petroleum refining: (1) removing carbon as coke, often referred to as carbon rejection, and producing lower-boiling products as overhead (distillate material); and (2) adding hydrogen by hydroprocessing by which the yield of coke is reduced in favor of lower boiling products (Figure 14.4).

The carbon rejection mechanism is, by definition, a means of removing a low-hydrogen product and allowing distribution of the inherent hydrogen among the lower boiling products. In addition, carbon rejection is, by definition, virtually a process by which the incompatibility of one product with the others is the means by which the process succeeds.

The first is the deasphalting process where asphaltenes,

and often resins, are discharged from the feedstock by the addition of hydrocarbon liquids.

This is analogous to the laboratory separation procedure (Mitchell and Speight, 1973), with the exception that the process liquids are often the lower-molecular-weight hydrocarbons liquefied under pressure (Speight, 1991). A similar situation occurs when asphaltenes are deposited on reservoir rock due to the increased solubility of hydrocarbon gases in the petroleum as reservoir pressure increases during maturation (Evans et al., 1971; Bailey et al., 1974).

The interesting point is that the amount of precipitate can be equated to the solubility parameter of the liquid phase (Mitchell and Speight, 1973) and offers the attractive option of predictability. In addition, the amount of precipitate varies with the amount of added paraffin liquid (Figure 14.5), which may be more representative of the situation that occurs in many reactors during refinery operations.

The effect of the higher temperatures on the separation of asphaltic material is largely unknown, although there are indications (Rao and Serrano, 1986) that higher temperatures discourage aggregation. In the high-temperature reactor, it must be presumed that it is the precursors to coke that separate from the liquid phase.

2.2 Visbreaking

As examples of processes in which compatibility might occur, it is appropriate, for the present purposes, to select the visbreaking process (i.e., a carbon rejection process) and the hydrocracking process (i.e., a hydrogen addition process), as would occur in an integrated refinery. The processes operate under different conditions (Figure 14.6) and have different levels of conversion (Figure 14.7).

The visbreaking process (Figure 14.8) is a means of reducing the viscosity of heavy feedstocks by "controlled" thermal decomposition (Speight, 1991). However, the process is often plagued by sediment formation in the products. This sediment, or sludge, must be removed if the products are to meet fuel oil specifications.

The process uses the approach of mild thermal cracking as a relatively low cost and low severity approach to improving the

357

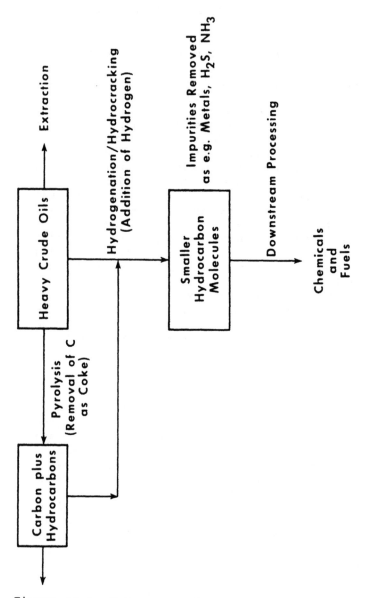

Figure 14.4: Hydroprocessing, Coke Yield and Lower Boiling
Products Yield.

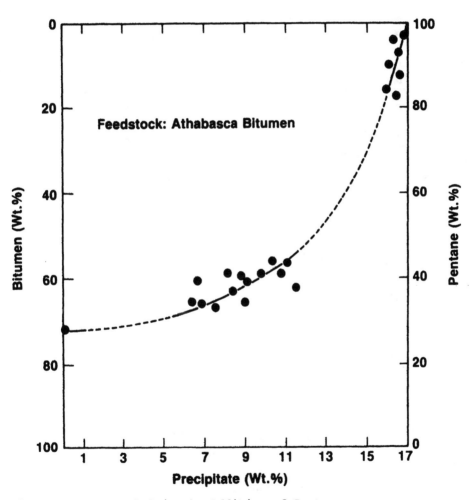

Figure 14.5: Deasphalting by Addition of Pentane.

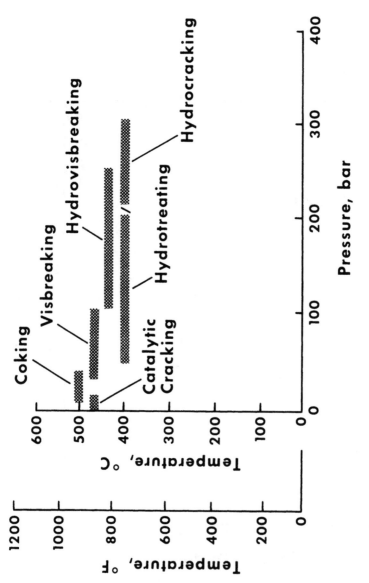

Figure 14.6: Processes That Operate in an Integrated Refinery.

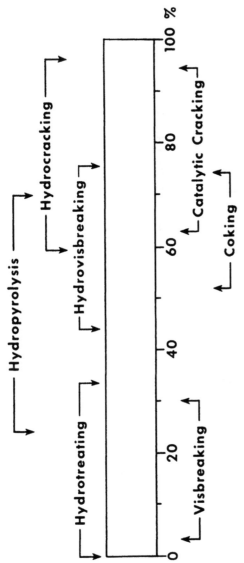

Figure 14.7: Level of Conversion in an Integrated Refinery.

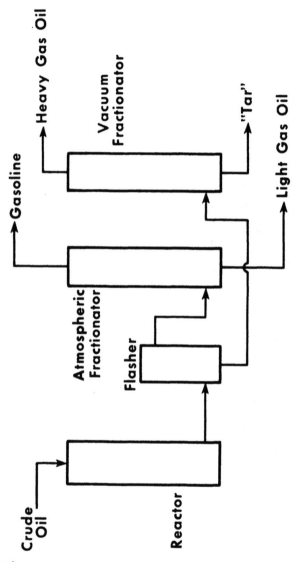

Figure 14.8: The Visbreaking Processes as a Method for Reducing Viscosity of Heavy Feedstocks.

viscosity characteristics of the residue without attempting significant conversion to distillates. Low residence times are required to avoid polymerization and coking reactions, although additives can help to suppress coke deposits on the tubes of the furnace (Allan et al., 1983).

The process consists of a reaction furnace, followed by quenching with recycled oil, and fractionation of the product mixture. All of the reactions in this process occur as the oil flows through the tubes of the reaction furnace. The severity is controlled by the flow rate through the furnace and the temperature; typical conditions are 475-500°C (885-930°F) at the furnace exit with a residence time of 1-3 min, with operation for 3-6 months before the furnace tubes must be cleaned and the coke removed (Gary and Handwerk, 1984). The operating pressure in the furnace tubes can range from 0.7 MPa to 5 MPa, depending on the degree of vaporization and the residence time desired. For a given furnace-tube volume, a lower operating pressure will reduce the actual residence time of the liquid phase.

The reduction in viscosity of the unconverted residue tends to reach a limiting value with conversion, although the total product viscosity can continue to decrease (Gray, 1994). Conversion of residue in visbreaking follows first-order reaction kinetics (Henderson and Weber, 1965). The minimum viscosity of the unconverted residue can lie outside the range of allowable conversion if sediment begins to form (Rhoe and de Blignieres, 1979). When pipelining of the visbreaker product is the process objective, addition of a diluent such as gas condensate can be used to achieve a further reduction in viscosity.

The high viscosity of the heavier feedstocks and residua is thought to be due to entanglement between the high-molecular-weight components of the oil and to the formation of ordered structures in the liquid phase. Thermal cracking at low conversion can remove side chains from the asphaltenes and break bridging aliphatic linkages. A 5-10% conversion of atmospheric residue to naphtha is sufficient to reduce the entanglements and structures in the liquid phase, and give at least a fivefold reduction in viscosity. Reduction in viscosity is also accompanied by a reduction in the pour point.

There are several chemical models (e.g., Schucker and

Keweshan, 1980; Gray, 1994 and references cited therein) that describe the thermal decomposition of asphaltenes and that can be described in composite form (Figure 14.9) and are worthy of consideration. Using the asphaltene models (Chapter 13) as a guide, the asphaltene nuclear fragments become progressively more polar as the paraffinic fragments are stripped from the ring systems by scission of the bonds (preferentially) between the carbon atoms alpha and beta and the aromatic rings (Figures 14.10 and 14.11).

The polynuclear aromatic systems that have been denuded of the attendant hydrocarbon moieties are somewhat less soluble (Bjorseth, 1983; Dias, 1987, 1988) in the surrounding hydrocarbon medium than their "parent" systems, not only because of the solubilizing alkyl moieties but also because of the enrichment of the liquid medium in paraffinic constituents. Again, there is an analogy to the deasphalting process, except the paraffinic material is a product of the thermal decomposition of the asphaltene molecules and is formed in situ rather than being added separately.

The stability of visbroken products is also an issue that might be addressed at this time. Using this simplified model, visbroken products might well contain such polar species that have been denuded of some of the alkyl chains and that, on the basis of solubility, might be more rightly called carbenes and carboids, but where an induction period is required for phase separation or agglomeration to occur.

Such products might, initially, be "soluble" in the liquid phase but after the induction period, cooling, and/or diffusion of the products, incompatibility (phase separation, sludge formation, agglomeration) occurs.

2.3 Coking

Coking, as the term is used in the petroleum industry, is a process for converting nondistillable fractions (residua) of crude oil to lower-boiling products and coke. Coking is often used in preference to catalytic cracking because of the presence of metals and nitrogen components that poison catalysts. There are actually several coking processes, i.e., delayed coking, fluid coking, and flexicoking, as well as several other variations.

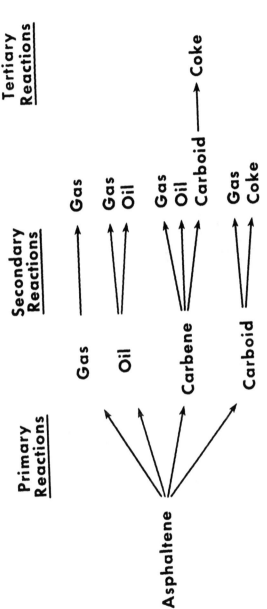

Figure 14.9: The Thermal Decomposition of Asphaltenes.

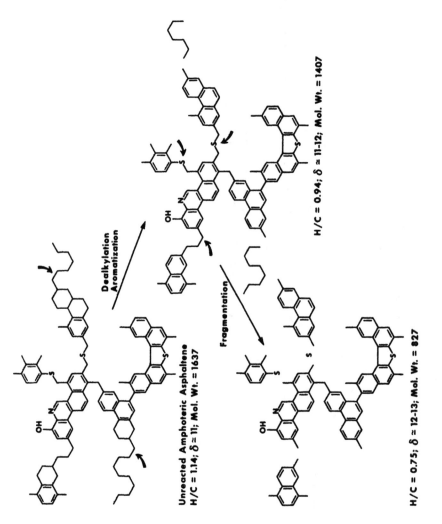

Figure 14.10: Model System for the Stripping of Paraffin Fragments From Asphaltenes.

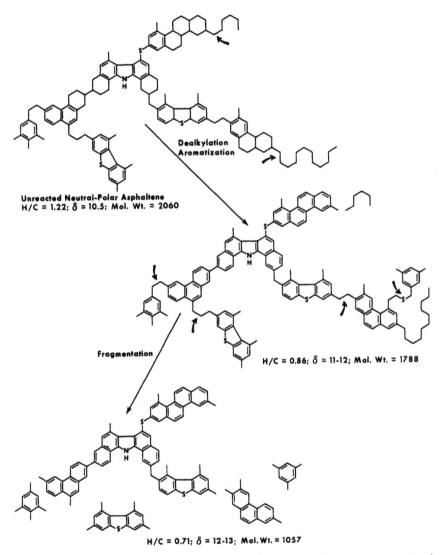

Unreacted Neutral-Polar Asphaltene
H/C = 1.22; δ = 10.5; Mol. Wt. = 2060

Dealkylation
Aromatization

Fragmentation

H/C = 0.86; δ = 11-12; Mol. Wt. = 1788

H/C = 0.71; δ = 12-13; Mol. Wt. = 1057

Figure 14.11: Model System for the Formation of Aromatic Fragments From Asphaltenes.

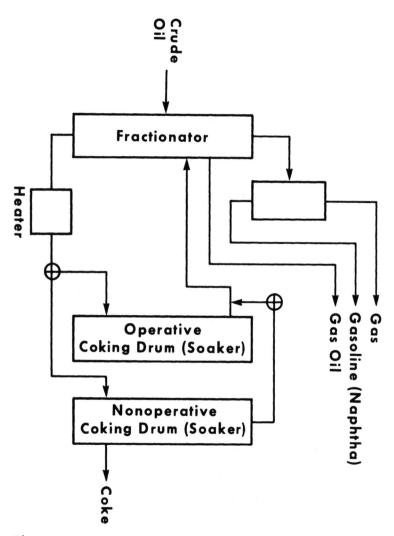

Figure 14.12: Delayed Coking Process.

Delayed coking (Figure 14.12) is the oldest, most widely used process and has changed very little in the five or more decades that it has been on-stream in refineries (Speight, 1991). In the semicontinuous process, the residuum or other heavy feedstock is heated to the cracking/coking temperature (>350°C, >660°F; but usually at temperatures of the order of 480°C, 895°F) of the feed, and the hot liquid is charged, usually by upflow, to the coke drum, where the coking reactions occur.

Liquid and gaseous products pass to the fractionator for separation, and coke is deposited in the drum. The coke drums are arranged in pairs, one on-stream and the other off-stream, and used alternately to allow for continuous processing on a cycle, typically 24-48 hr.

The overhead oil is fractionated into fuel gas (ethane and lower-molecular-weight gases), propane-propylene, butane-butene, naphtha, light gas oil, and heavy gas oil. Yields and product quality vary widely due to the broad range of feedstock types charged to delayed coking. The function of the coke drum is to provide the residence time required for the coking reactions and to accumulate the coke. Hydraulic cutters are used to remove coke from the drum.

Fluid coking (Figure 14.13) is a continuous fluidized solids process that cracks feed thermally over heated coke particles in a reactor vessel into gas, liquid products, and coke. Heat for the process is supplied by partial combustion of the coke, with the remaining coke being drawn as product. The new coke is deposited in a thin fresh layer on the outside surface of the circulating coke particle.

Small particles of coke made in the process circulate in a fluidized state between the vessels and are the heat transfer medium. Thus, the process requires no high-temperature preheat of the furnace. Fluid coking is carried out at essentially atmospheric pressure and temperatures in excess of 485°C (900°F), with residence times of the order of 15-30 sec. The longer residence time is in direct contrast to the delayed coking process, where the coking reactions are allowed to proceed to completion. This is evident from the somewhat higher liquid yields observed in many fluid coking processes. However, the products from a fluid coker may be somewhat more olefinic

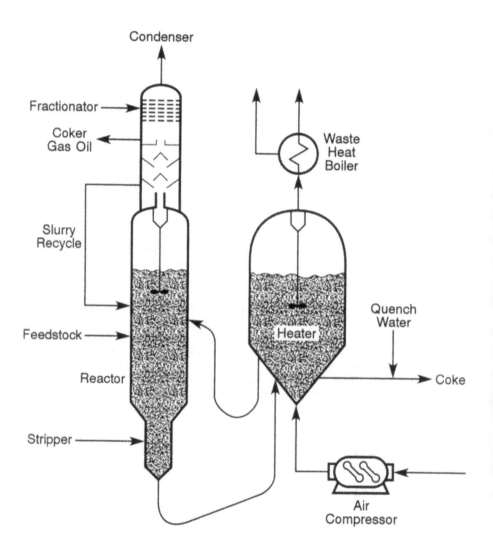

Figure 14.13: Fluid Coking Process.

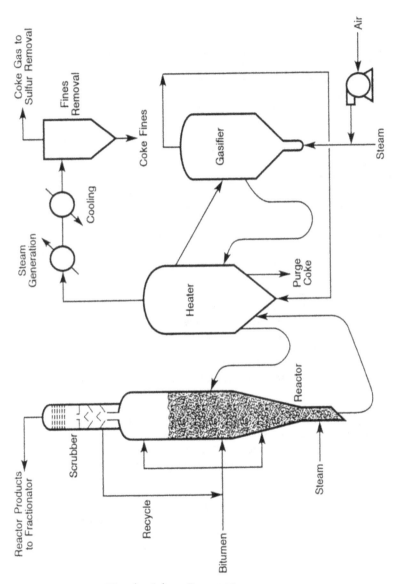

Figure 14.14: Flexicoking Process.

and slightly less desirable for downstream processing.

The **flexicoking** (Figure 14.14) process is a modification of the fluid coking process insofar as the process also includes a gasifier adjoining the burner/regenerator to convert excess coke to a clean fuel gas with a heating value of about 90 Btu/ft^3. The coke gasification can be controlled to burn about 95% of the coke to maximize production of coke gas or at a reduced level to produce both gas and a coke. This flexibility permits adjustment for coke market conditions over a considerable range of feedstock properties.

Suitable feedstocks include vacuum residua of all types, asphalt, tar sand bitumen, and visbreaker bottoms.

The **liquid products** from the coker can, following cleanup via commercially available hydrodesulfurization technology (Speight, 1981), provide low-sulfur liquid fuels (less than 0.2% w/w sulfur). Coker naphthas have boiling ranges up to 220°C (430°F), are olefinic, and must be upgraded by hydrogen processing for removal of olefins, sulfur, and nitrogen. They are then used conventionally for reforming to gasoline or chemicals feedstock. Middle distillates, boiling in the range 220-360°C (430-680°F), also are hydrogen treated for improved storage stability, sulfur removal, and nitrogen reduction. They can then be used as precursors to gasoline, diesel fuel, or fuel oil. The gas-oil boiling up to 510°C (950°F) is usually low in metals and may be used as the feedstock for fluid catalytic cracking.

Another major application for the coking processes is upgrading heavy low-value crude oils into lighter products.

Petroleum **coke** is used principally as a fuel or, after calcining, for carbon electrodes. The feedstock from which the coke is produced controls the coke properties, especially in terms of sulfur, nitrogen, and metals content. A concentration effect tends to deposit the majority of the sulfur, nitrogen, and metals in the coke. Cokes exceeding ca. 2.5% sulfur content and 200 ppm vanadium are mainly used, environmental regulations permitting, for fuel or fuel additives. The properties of coke necessary for nonfuel use include low sulfur, metals, and ash as well as a definable physical structure.

2.4 Hydrotreating/Hydrocracking

In contrast to the visbreaking process, in which the general principle is the production of products for use as fuel oil, the hydrocracking process is employed to produce a slate of products for use as liquid fuels. Nevertheless, the decomposition of asphaltenes is, again, an issue, and just as models consisting of large polynuclear aromatic systems are inadequate to explain the chemistry of visbreaking, they are also of little value for explaining the chemistry of hydrocracking.

The presence of hydrogen changes the nature of the products, especially the coke yield, by preventing the buildup of precursors that are incompatible in the liquid medium and form coke (Magaril and Aksenova, 1968; Magaril and Ramazaeva, 1969; Magaril et al., 1970). In fact, the nearly impossible chemistry involved in the reduction of asphaltenes to liquids using models where the polynuclear aromatic system borders on graphitic is difficult to visualize. However, the "paper" chemistry derived from the use of a model composed of smaller polynuclear aromatic systems is much easier to visualize (Speight, 1991). Precisely how asphaltenes react with the catalysts is open to much more speculation.

Incompatibility is still possible when asphaltenes interact with catalysts, especially acidic support catalysts, through the functional groups, e.g., the basic nitrogen species, just as they interact with adsorbents. There is also the possibility for interaction of the asphaltene with the catalyst through the agency of a single functional group in which the remainder of the asphaltene molecule remains in the liquid phase. There is also a less desirable, option in which the asphaltene reacts with the catalyst at several points of contact, causing immediate incompatibility on the catalyst surface.

2.5 Asphalt Manufacture

Another area where incompatibility might play a detrimental role, during processing or in the product, is in asphalt oxidation.

The more polar species in a feedstock will oxidize first during air blowing. After incorporation of oxygen to a limit, significant changes can occur in the asphaltene structure (Figure 14.15), especially in terms of the incorporation of

373

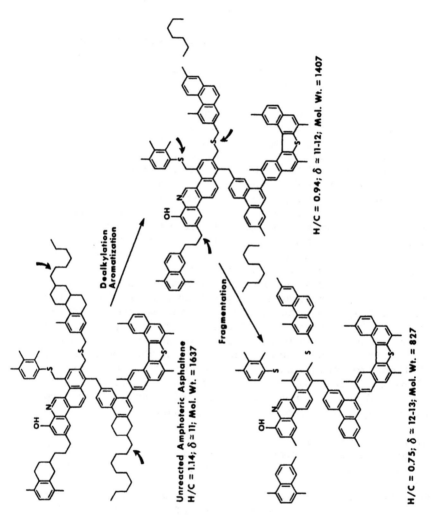

Figure 14.15: Model Asphaltene Chemical Structure.

polar oxygen, which can influence molecular weight. Thus, the change in the character of the asphalt may not be so much due to oxidative degradation but to the incorporation of oxygen functions that interfere with the natural order of intramolecular structuring.

There is the potential that the incorporation of oxygen functions enhances the ability of the asphalt to bond to the aggregate. Uncontrolled introduction of oxygen functions could result in the production of a poor-grade asphalt, where phase separation of the oxidized asphaltene may already have occurred, or, should it occur in the product, the result can be pavement failure due to a weakening of the asphalt-aggregate interactions.

3.0 Process Chemistry and Physics

When petroleum fractions are heated to temperatures over ca. 410°C (770°F), the thermal or free radical reactions start to give way to cracking of the mixture at significant rates.

Thermal conversion does not require the addition of a catalyst; therefore, this approach is the oldest technology available for residue conversion. The severity of thermal processing determines the conversion and the product characteristics.

Thermal treatment of residua ranges from mild treatment for reduction of viscosity to extremely high temperatures (>1000°C, >1830°F) for complete conversion to olefins and light ends. The higher the temperature, the shorter the time to achieve a given conversion. The severity of the process conditions is the combination of reaction time and temperature to achieve a given conversion.

If no side reactions occur, then very long times at low temperatures should be equivalent to very short times at high temperatures. Thermal reactions, however, can give rise to a variety of different reactions, so that selectivity for a given product changes with temperature and pressure. In fact, the nature of the thermal processes varies considerably.

The mild- and high-severity processes are frequently used for processing of residue fractions, while conditions similar to ultrapyrolysis (extremely high temperature pyrolysis)" (high temperature and very short residence time) are only used commercially for cracking ethane, propane, butane, and light distillate feeds to produce ethylene and higher olefins.

The available data suggest that coke formation is a complex process involving both chemical reactions and thermodynamic behavior. Like the asphaltenes, coke should be viewed as a solubility fraction. Its physical state at room temperature is solid, and it is insoluble in benzene or toluene.

The following two-step mechanism has emerged from studies of whole oils and solubility fractions (Savage et al., 1988; Wiehe, 1993a, 1993b):

(1) Thermal reactions result in the formation of high-molecular-weight, aromatic components in solution in the liquid phase.

(2) Once the concentration of this material reaches a critical value, phase separation occurs, giving a denser, aromatic liquid phase.

Reactions that contribute to this process are cracking of side chains from aromatic groups, dehydrogenation of naphthenes to form aromatics, condensation of aliphatic structures to form aromatics, condensation of aromatics to form higher fused-ring aromatics, and dimerization of oligomerization reactions. Loss of side chains always accompanies thermal cracking, while dehydrogenation and condensation reactions are favored by hydrogen deficient conditions.

The importance of solvents in coking has been recognized for many years, but their effects have often been ascribed to hydrogen-donor reactions rather than phase behavior. The separation of the phases depends on the solvent characteristics of the liquid. Addition of aromatic solvents will suppress phase separation, while paraffins will enhance separation. Microscopic examination of coke particles often shows evidence for mesophase, spherical domains that exhibit the anisotropic optical characteristics of liquid crystal.

This phenomenon is consistent with the formation of a second liquid phase; the mesophase liquid is denser than the rest of the hydrocarbon, has a higher surface tension, and likely wets metal surfaces better than the rest of the liquid phase. The mesophase characteristic of coke diminishes as the liquid phase becomes more compatible with the aromatic material.

A simple kinetic model has been proposed to account for the data (Wiehe, 1993a, 1993b). Thus,

$$M \text{---} k_m \text{---}> a A* + (1 - a)V$$
$$A \text{---} k_a \text{---}> b A* + c M* + (1 - b - c)V$$
$$A*_{max} = S_1 (M + M*)$$
$$A*_{ex} = A* - A*_{max}$$

At long reaction times,

$$A*_{ex} \text{-----}> y C + (1 - y)M*$$

where M and A are maltenes and asphaltenes in the feed, M* and A* are the corresponding products that may have undergone change from the original materials, V is volatile matter, and C is coke. The parameters a, b, c, and y are stoichiometric coefficients. The solubility limit, S_1, is given as a fraction of total maltenes, and this limit determines the maximum cracked asphaltene concentration in solution, $A*_{max}$. The excess asphaltene, $A*_{ex}$, separates as a second phase and gives way to coke formation.

The ability of such a model to correlate data from pyrolysis of maltenes and whole residue with a single set of stoichiometric parameters suggests that this semi-empirical approach has significant merit. The model was consistent with the observed induction behavior, and the maximum in asphaltene concentration. The solubility limit, S_1, depended on the initial composition, as expected for a thermodynamic property of the reaction mixture. A more complete model will require prediction of the solubility limit based on the properties of the maltenes (solvent) and the reacted asphaltenes (solute), possibly using the solubility parameter.

Most incompatibility phenomena can also be explained by the use of the solubility parameter, δ, for petroleum fractions and for the solvents. Although little is known about the solubility parameter of petroleum fractions, there has been a noteworthy attempt to define the solubility parameter ranges for different fossil fuel liquids (Figure 14.16) (Yen, 1984).

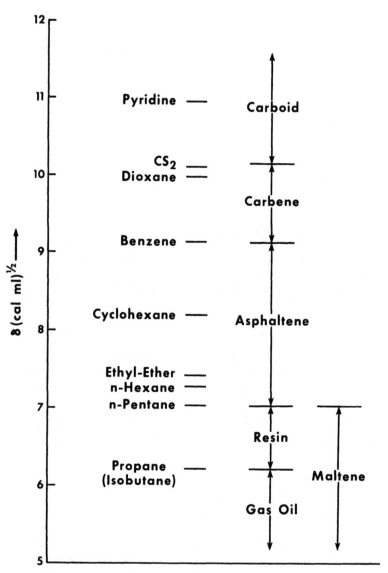

Figure 14.16: Solubility Parameter Ranges for Fossil Fuel Liquids.

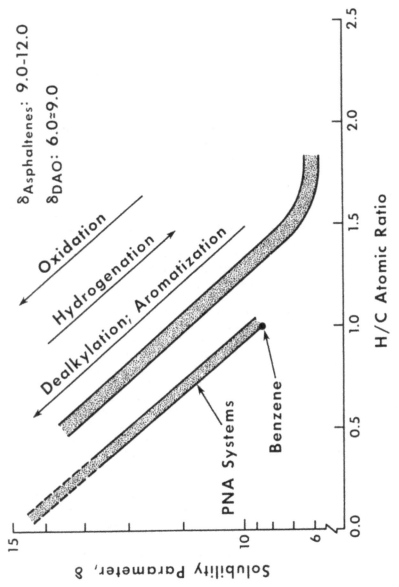

Figure 14.17: Reactions of Asphaltenes and Changes in Solubility
Parameter.

As an extension of this concept, there is sufficient fragmented data to draw an approximate correlation between H/C atomic ratio and δ for hydrocarbons and the constituents of the lower-boiling fractions of petroleum. Note that hydrocarbon liquids can dissolve polynuclear hydrocarbons where there is, usually, less than a three-point difference between the lower solubility parameter of the solvent and the higher solubility parameter of the solute. Thus, a parallel, or near-parallel, line can be assumed that allows the solubility parameter of the asphaltenes and resins to be estimated.

By this means, the solubility parameter of asphaltenes can be estimated to fall in the range 9-12, which is also in keeping with the asphaltenes being composed of a mixture of different compound types with the accompanying variation in polarity. Removal of alkyl side chains from the asphaltenes will decrease the H/C ratio (Speight, 1991; Wiehe, 1993a, 1993b; Gray, 1994) and increase the solubility parameter, thereby bringing about a concurrent decrease of the asphaltene product in the hydrocarbon solvent.

In fact, on the molecular weight/polarity diagram for asphaltenes, carbenes and carboids can be shown as lower-molecular-weight, highly polar entities, in keeping with molecular fragmentation models (Figures 14.10 and 14.11).

In terms of asphalt oxidation and ensuing incompatibility, the asphaltenes will oxidize, which in turn, will cause an increase in their solubility parameter (Figure 14.17). The soloubility parameter of petroleum fractions can be estimated from the structural types present on the basis of a group contribution method. This is an area where model structures for asphaltenes (Chapter 13) can be used constructively and gainfully. The data offer the beginnings of an understanding of asphaltene behavior, at the molecular level.

As with the cracked products, if this increase in polarity and solubility parameter is too drastic relative to the surrounding medium, phase separation will occur immediately or over a period of time, depending upon the extent of the oxidation. Once phase separation occurs, instability of the asphalt-aggregate mix ensues.

The use of the models can give an understanding of the chemical and physical aspects of incompatibility. Thus, it is possible to explain refining chemistry and physics, to some extent, by the use of molecular models as an illustration of how incompatibility might originate. Indeed, it is also possible to use the models to explain certain aspects of reservoir and recovery chemistry and physics.

The models presented differ from previous models insofar as they include polynuclear aromatic systems that are more compatible with the other fractions of petroleum and are based upon the natural product origins of oil. Other models will undoubtedly be developed that seem just as, or perhaps even more, appropriate than those given here. However, it should never be forgotten that such models must be a "means to an end," allowing researchers to explain feedstock behavior and affording the luxury of "understandability."

Finally, although the focus in this work has been on the heavier feedstocks, similar principles might also be applied to incompatibility as evidenced in various fuel oil products (Chapters 8, 9, and 10).

Any chemical or physical interaction that causes a change in the solubility parameter of the solute relative to that of the solvent will also cause incompatibility, be it called instability, incompatibility, phase separation, or sediment/sludge formation.

PETROLEUM PRODUCTS: INSTABILITY AND INCOMPATIBILITY

4.0 References

Allan, D.E., Martinez, C.H., Eng. C.C., and Barton, W.J. 1983. Chem. Eng. Progr. 79(1): 85.

Bailey, M.J.L., Evans, C.R., and Milner, C.W.D. 1974. Bull. Am. Assoc. Pet. Geol.: 58: 2284.

Bjorseth, A. 1983. Handbook of Polycyclic Aromatic Hydrocarbons. Marcel Dekker Inc., New York.

Dias, J.R. 1987. Handbook of Polycyclic Hydrocarbons. Part A. Benzenoid Hydrocarbons. Elsevier, New York.

Dias, J.R. 1988. Handbook of Polycyclic Hydrocarbons. Part B. Polycyclic Isomers and Analogs of Benzenoid Hydrocarbons. Elsevier, New York.

Evans, C.R., Rogers, M.A., and Bailey, M.J.L. 1971. Chem. Geol. 8: 147.

Gary, J.G., and Handwerk, G.E. 1984. Petroleum Refining: Technology and Economics. Marcel Dekker Inc., New York.

Gray, M.R. 1994. Upgrading Petroleum Residues and Heavy Oils. Marcel Dekker Inc., New York.

Henderson, J.H., and Weber, L. 1965. J. Can. Pet. Technol. 4: 206.

Magaril, R.Z., and Aksenova, E.I. 1968. Int. Chem. Eng. 8: 727.

Magaril, R.Z., and Ramazaeva, L.F. 1969. Izv. Vyssh. Uchebn. Zaved Neft Gaz. 12(1): 61.

Magaril, R.Z., Ramazaeva, L.F., and Aksenova, E.I. 1970. Khim. Tekhnol. Topliv. Masel. 15(3): 15.

Mitchell, D.L., and Speight, J.G. 1973. Fuel. 52: 149.

Rao, B.M.L., and Serrano, J.E. 1986. Fuel Sci. and Technol. Int. 4: 483.

Rhoe, A., and de Blignieres, C. 1979. Hydrocarbon Process. 58(1): 131.

Savage, P.E., and Klein, M.T. 1989. Chem. Eng. Sci. 44: 393.

Savage, P.E., Klein, M.T., and Kukes, S.G. 1988. Energy Fuels. 2: 619.

Schucker, R.C., and Keweshan, C.F. 1980. Preprints Div. Fuel Chem. Am. Chem. Soc. 25: 155.

Speight, J.G. 1981. The Desulfurization of Heavy Oils and Residua. Marcel Dekker Inc., New York.

Speight, J.G. 1987. Preprints. Div. Petrol. Chem. Am. Chem. Soc. 32(2): 413.

Speight, J.G. 1991. The Chemistry and Technology of Petroleum. 2nd Edition. Marcel Dekker Inc., New York.

Speight, J.G. 1992. Proceedings, 4th International Conference on the Stability and Handling of Liquid Fuels. Report DOE/CONF-911102. U.S. Department of Energy, Washington, DC. P. 169.

Speight, J.G., and Moschopedis, S.E. 1979. Fuel Process. Technol. 2: 295.

Wiehe, I.A. 1992. Ind. Eng. Chem. Res. 31: 530.

Wiehe, I.A. 1993a. Ind. Eng. Chem. Res. 32: 2447.

Wiehe, I.A. 1993b. Preprints Div. Pet. Chem. Am. Chem. Soc. 38: 428.

Yen, T.F. 1984. In The Future of Heavy Crude Oil and Tar Sands. R.F. Meyer, J.C. Wynn, and J.C. Olson (editors). McGraw-Hill, New York.

390